Pioneers, Leaders and Followers in Multilevel and Polycentric Climate Governance

Pioneers, Leaders and Followers in Multilevel and Polycentric Climate Governance focuses on pioneers, leaders and followers as central drivers for international climate change governance innovations.

A burgeoning literature has identified a wide range of actors, such as international organisations, the European Union, NGOs, corporations and cities, as potential and actual climate pioneers and/or leaders in international climate change. Despite this, much of the academic debate is still largely focused on states. To address this research gap, this volume focuses primarily on non-state actors in different multilevel and polycentric governance structures. The chapters offer a critical analysis of the different types of actors (e.g. the EU, corporate actors, NGOs and cities) who can act as pioneers and/or leaders at different levels of climate governance (including the international, supranational, regional, national and local) encompassing non-state and state actors. The volume provides a clear conceptualisation of pioneers, leaders and followers while assessing their motives, capacities, styles and strategies. It examines critically the dynamic interrelationship between leaders and pioneers on the one hand, and followers and laggards on the other. Moreover, it analyses how multilevel and polycentric climate governance structures enable and/or constrain climate pioneers, leaders and followers.

This volume will be of great use to scholars of environmental governance, climate change and international governance.

The chapters were originally published as a special issue in *Environmental Politics*.

Rüdiger K.W. Wurzel is a Professor of Comparative European Politics and Jean Monnet Chair in European Union Studies at the University of Hull, UK, where he is the Director of the Centre for European Union Studies (CEUS).

Duncan Liefferink is an Assistant Professor in the Environmental Governance and Politics Group, Institute for Management Research at Radboud University Nijmegen, The Netherlands.

Diarmuid Torney is an Associate Professor in the School of Law and Government at Dublin City University, Ireland.

Pioneers, Leaders and Followers in Multilevel and Polycentric Climate Governance

Edited by
Rüdiger K.W. Wurzel, Duncan Liefferink
and Diarmuid Torney

LONDON AND NEW YORK

First published 2020
by Routledge
2 Park Square, Milton Park, Abingdon, Oxon, OX14 4RN

and by Routledge
52 Vanderbilt Avenue, New York, NY 10017

Routledge is an imprint of the Taylor & Francis Group, an informa business

© 2020 Taylor & Francis

Chapters 5 & 7 © 2017 Matthew Bach & Kristine Kern. Originally published as Open Access.

With the exception of Chapters 5 and 7, no part of this book may be reprinted or reproduced or utilised in any form or by any electronic, mechanical, or other means, now known or hereafter invented, including photocopying and recording, or in any information storage or retrieval system, without permission in writing from the publishers. For details on the rights for Chapters 5 and 7, please see the chapters' Open Access footnotes.

Trademark notice: Product or corporate names may be trademarks or registered trademarks, and are used only for identification and explanation without intent to infringe.

British Library Cataloguing-in-Publication Data
A catalogue record for this book is available from the British Library

ISBN13: 978-0-367-46759-3

Typeset in Minion Pro
by codeMantra

Publisher's Note
The publisher accepts responsibility for any inconsistencies that may have arisen during the conversion of this book from journal articles to book chapters, namely the inclusion of journal terminology.

Disclaimer
Every effort has been made to contact copyright holders for their permission to reprint material in this book. The publishers would be grateful to hear from any copyright holder who is not here acknowledged and will undertake to rectify any errors or omissions in future editions of this book.

Contents

Citation Information vii
Notes on Contributors ix

1 Pioneers, leaders and followers in multilevel and polycentric
  climate governance 1
  Rüdiger K.W. Wurzel, Duncan Liefferink and Diarmuid Torney

2 Leadership and lesson-drawing in the European Union's
  multilevel climate governance system 22
  Martin Jänicke and Rüdiger K.W. Wurzel

3 Environmental, climate and social leadership of small
  enterprises: Fairphone's step-by-step approach 43
  Katja Biedenkopf, Sarah Van Eynde and Kris Bachus

4 Climate pushers or symbolic leaders? The limits to corporate
  climate leadership by food retailers 64
  Johann Dupuis and Remi Schweizer

5 The oil and gas sector: from climate laggard to climate leader? 87
  Matthew Bach

6 Oil and power industries' responses to EU emissions trading:
  laggards or low-carbon leaders? 104
  Per Ove Eikeland and Jon Birger Skjærseth

7 Cities as leaders in EU multilevel climate governance: embedded
  upscaling of local experiments in Europe 125
  Kristine Kern

8 Climate pioneership and leadership in structurally disadvantaged
  maritime port cities 146
  *Rüdiger K.W. Wurzel, Jeremy F.G. Moulton, Winfried Osthorst,
  Linda Mederake, Pauline Deutz and Andrew E.G. Jonas*

9 Follow the leader? Conceptualising the relationship between
  leaders and followers in polycentric climate governance   167
  Diarmuid Torney

Index   187

Citation Information

The chapters in this book were originally published in *Environmental Politics*, volume 28, issue 1 (January 2019). When citing this material, please use the original page numbering for each article, as follows:

Chapter 1
Pioneers, leaders and followers in multilevel and polycentric climate governance
Rüdiger K.W. Wurzel, Duncan Liefferink and Diarmuid Torney
Environmental Politics, volume 28, issue 1 (January 2019) pp. 1–21

Chapter 2
Leadership and lesson-drawing in the European Union's multilevel climate governance system
Martin Jänicke and Rüdiger K.W. Wurzel
Environmental Politics, volume 28, issue 1 (January 2019) pp. 22–42

Chapter 3
Environmental, climate and social leadership of small enterprises: Fairphone's step-by-step approach
Katja Biedenkopf, Sarah Van Eynde and Kris Bachus
Environmental Politics, volume 28, issue 1 (January 2019) pp. 43–63

Chapter 4
Climate pushers or symbolic leaders? The limits to corporate climate leadership by food retailers
Johann Dupuis and Remi Schweizer
Environmental Politics, volume 28, issue 1 (January 2019) pp. 64–86

Chapter 5
The oil and gas sector: from climate laggard to climate leader?
Matthew Bach
Environmental Politics, volume 28, issue 1 (January 2019) pp. 87–103

Chapter 6
Oil and power industries' responses to EU emissions trading: laggards or low-carbon leaders?
Per Ove Eikeland and Jon Birger Skjærseth
Environmental Politics, volume 28, issue 1 (January 2019) pp. 104–124

Chapter 7
Cities as leaders in EU multilevel climate governance: embedded upscaling of local experiments in Europe
Kristine Kern
Environmental Politics, volume 28, issue 1 (January 2019) pp. 125–145

Chapter 8
Climate pioneership and leadership in structurally disadvantaged maritime port cities
Rüdiger K.W. Wurzel, Jeremy F.G. Moulton, Winfried Osthorst, Linda Mederake, Pauline Deutz and Andrew E.G. Jonas
Environmental Politics, volume 28, issue 1 (January 2019) pp. 146–166

Chapter 9
Follow the leader? Conceptualising the relationship between leaders and followers in polycentric climate governance
Diarmuid Torney
Environmental Politics, volume 28, issue 1 (January 2019) pp. 167–186

For any permission-related enquiries please visit:
http://www.tandfonline.com/page/help/permissions

Contributors

Matthew Bach works as a Governance Programme Officer at ICLEI European Secretariat - Local Governments for Sustainability.

Kris Bachus is a Research Manager of climate and sustainability at the HIVA Research Institute, KU Leuven, Belgium.

Katja Biedenkopf is an Associate Professor of Sustainability Politics at Leuven International and European Studies (LINES), KU Leuven, Belgium.

Jon Birger Skjærseth is a Research Professor at the Fridtjof Nansen Institute, Norway.

Pauline Deutz is a Reader in the Department of Geography, Geology and Environment at the University of Hull, UK.

Johann Dupuis is a Senior Researcher in the Swiss Graduate School of Public Administration (IDHEAP) at the University of Lausanne, Switzerland.

Per Ove Eikeland is a Senior Research Fellow at the Fridtjof Nansen Institute, Norway.

Martin Jänicke has 40 years of experience as professor for comparative politics, scientific author and senior policy advisor.

Andrew E.G. Jonas is a Professor of human geography at the University of Hull, UK.

Kristine Kern is a Senior Researcher at the Leibniz Institute for Research on Society and Space (IRS), Erkner, Germany and an Adjunct Professor at Åbo Akademi University, Turku, Finland.

Duncan Liefferink is an Assistant Professor in the Environmental Governance and Politics Group, Institute for Management Research at Radboud University Nijmegen, The Netherlands.

Linda Mederake works as a Researcher at Ecologic Institute in Berlin, Germany.

Jeremy F.G. Moulton is an Associate Lecturer in the Department of Environment & Geography at the University of York, UK.

Winfried Osthorst is a Professor of Governance in Multilevel Systems and Global Change in the Faculty of Social Sciences at the University of Applied Sciences Bremen, Germany.

Remi Schweizer is a Project Manager at the Sustainable Development Unit for the Vaud State, Switzerland.

Diarmuid Torney is an Associate Professor in the School of Law and Government in Dublin City University, Ireland.

Sarah Van Eynde is a PhD Researcher at Leuven International and European Studies (LINES), KU Leuven, Belgium.

Rüdiger K.W. Wurzel is a Professor of Comparative European Politics and Jean Monnet Chair in European Union Studies at the University of Hull, UK, where he is the Director of the Centre for European Union Studies (CEUS).

Pioneers, leaders and followers in multilevel and polycentric climate governance

Rüdiger K.W. Wurzel ⓘ, Duncan Liefferink ⓘ and Diarmuid Torney ⓘ

Introduction

The environmental governance literature has seen a proliferation of analytical terms to describe actors who try to engender change for the improvement of the environment/climate, such as entrepreneur, forerunner, front runner, first mover, leader, lead state, pace setter, pioneer and trend setter (Liefferink and Wurzel 2017, for the general leadership literature see Rhodes and t'Hart 2014, p.3). This proliferation of concepts has inhibited the development of cumulative, theory-guided research on environmental and climate governance. Scholars widely see leaders and pioneers, which are the most commonly used terms in this burgeoning literature, as agents of change who are of central importance for climate change mitigation and adaptation (Liefferink and Wurzel 2018, p. 135). However, the precise ways in which different types of leaders and pioneers act and interact with followers in multilevel and polycentric climate governance structures have attracted much less scholarly attention (but see e.g. Torney 2015).

This introduction and the other contributions to this volume take as an analytical starting point the differentiation that leaders usually actively seek to attract followers while this is not normally the case for pioneers (Liefferink and Wurzel 2017). Although not primarily interested in attracting followers, others nevertheless may emulate pioneers. The actual impact of leaders and pioneers, which needs to be established empirically, is dependent on both their own actions and the ensuing dynamics or stalemates with followers and laggards.

Early leadership studies focused on the actions, strategies and motives of powerful individuals such as American Presidents (e.g. Burns 1978). The international relations (IR) environmental leadership literature shifted the focus to lead *states* and their actions in international negotiations (e.g.

Young 1991, Underdal 1994). Some scholars paid early attention also to *international organisations* (IOs) and the *supranational European Union* (EU) (e.g. Rehbinder and Stewart 1985) and *subnational actors* (e.g. Freeman 1996). The more specialised climate governance literature quickly took on a multi-actor perspective that focused not only on *states* but also on the *EU* (e.g. Grubb and Gupta 2000, Wurzel et al. 2017), *businesses* (e.g. Grant 2011), *NGOs* (e.g. Long et al. 2002, Bäckstrand et al. 2017), *unions* (e.g. Räthzel and Uzzell 2013) and even *individuals* (e.g. Rowlands 1995). However, scholars have conducted little systematic research on subnational and non-state actors as *leaders* or *pioneers*. This volume aims to make a contribution towards closing this research gap.

Multilevel governance and polycentricity

Contemporary climate governance scholarship is paying increasing attention to a wider range of governance actors and levels, including the international, supranational, transnational, national and subnational levels. Scholars have widely used multi-level governance (MLG) and polycentric governance perspectives to analyse climate governance (e.g. Jordan et al. 2018). The shift away from a top-down climate governance approach embodied in the 1997 Kyoto Protocol partly has driven this development. While the Protocol stipulates legally binding targets and timetables, the 2015 Paris Agreement, relies instead on voluntary pledges (so-called Nationally Determined Contributions). According to Oberthür (2016, p. 81), the Paris 'Agreement recalibrates the role of the multilateral UN process as providing overall direction towards global decarbonisation, while leaving implementation to other international organisations, states and various non-state actors and initiatives'. Similarly, Jordan et al. (2018, p. 4) have argued that developments in climate change governance since the 2000s 'appear to confirm the trend towards greater polycentricity' which requires the mobilisation of a wider range of non-state actors.

Polycentric governance concepts share certain core presuppositions (e.g. multiple centres of authority and levels of governance) with MLG approaches, although conceptually they are *not* identical (e.g. Homsy and Warner 2014, Wurzel et al. 2017, Jordan et al. 2018). By comparison with polycentric governance approaches, MLG concepts normally assume a stronger role for governmental (i.e. state, supranational and subnational) actors (Morrison et al. 2017, Liefferink and Wurzel 2018). Most MLG-inspired EU studies – for which the MLG concept was initially developed (e.g. Marks 1993, Hooghe 1996) – emphasise the mutual dependency of supranational and subnational governmental actors. Many MLG concepts reject the idea of traditional top-down *government* in favour of less hierarchical *governance* which assumes that 'non-state actors co-govern along with state actors for the provision of collective goods and adopt governance functions that have formerly been the sole authority of sovereign states' (Stephenson 2013, p. 829).

Polycentric governance concepts, on the other hand, attribute a high degree of autonomy to societal actors (e.g. business, NGOs and individual citizens). Polycentric approaches claim that widespread self-coordination leads to a multitude of decision-making 'centres' particularly at subnational level or even at the level of firms (see Bach 2019, Biedenkopf *et al.* 2019, Dupius and Schweizer 2019, Eikeland and Skjaerseth 2019, – all this volume). From a polycentric governance perspective, such self-coordination, which we could conceptualise as leadership/pioneership by societal actors within relatively autonomous policy domains, is essential for the successful functioning of global climate governance (e.g. Ostrom *et al.* 2012, Ostrom 2014, Jordan *et al.* 2018). According to Dorsch and Flachsland (2017), one of the advantages of polycentric governance is that experimentation at local and decentralised levels may lead to learning-by-doing and subsequent horizontal diffusion or upscaling to higher climate governance levels (see Ostrom *et al.* 2012, 2014, Kern 2019, – this volume).

Broadly speaking, we argue that proponents of polycentricity favour societal self-coordination within market-like governance structures (e.g. Ostrom *et al.* 2012, Ostrom 2014) while MLG advocates support the creation of networks in which governmental actors (including supranational EU actors) play an important, if not dominant, role to correct negative market externalities (e.g. Marks 1993, Hooghe 1996, Homsy and Warner 2014). This also has consequences for the conceptualisation of leadership and pioneership. A polycentric governance perspective in principle could see any successful self-coordination or experimentation at any governance level as *pioneership*. Polycentricity can help us understand why and how such initiatives emerge and flourish (Dorsch and Flachsland 2017, Jordan *et al.* 2018). However, due to the relative autonomy of polycentric sub-systems, the number of potential followers may be limited (Liefferink and Wurzel 2018). For the wider leader/pioneer–follower dynamic, the overarching MLG context in which polycentric systems may be embedded is important (Morrison *et al.* 2017).

Within the broader context of MLG and polycentric approaches, we provide a critical assessment of the climate leader and pioneer literature. We do so by proposing answers to the following four core research questions: *Who* can be a leader/pioneer? *Why* do actors become leaders/pioneers? *How* do leaders/pioneers act? and How do leaders/pioneers attract *followers*? Earlier studies have addressed some of these questions, but not in a specifically MLG and polycentric governance context. In particular, by focusing on the EU, business and subnational actors (including cities) as potential leaders and pioneers, the contributions to this volume provide new answers to the *who*, *why* and *how* questions of leadership and pioneership. Moreover, the focus on followership opens a new dimension to the leadership/pioneership literature. By focusing on agents of change (i.e. leaders and pioneers), we draw heavily on actor-focused approaches (which are compatible with MLG and polycentric governance concepts) although we do not ignore structural factors (e.g. structural leadership).

Who can be a leader or a pioneer?

Although states, IOs, the supranational EU, subnational actors and societal actors have all been identified as potential environmental leaders or pioneers, the early climate governance literature focused primarily on states. However, competing conceptual approaches have emphasised the relative importance of particular types of actors who, as we explain below, may offer different types of leadership/pioneership and followership.

State-centred approaches still dominate both IR and comparative politics (CP) research on environmental/climate leaders and pioneers. The early IR literature focused on the leading role of states in international environmental/climate *regimes* (e.g. Young 1991, Underdal 1994) while CP studies have assessed the political characteristics and institutional capacities of states that transform into environmental leaders/pioneers (e.g. Jänicke and Weidner 1997, Weidner et al. 2002). In contrast to international regimes, institutionalist approaches have seen *IOs* as being capable of exhibiting independent 'actorness' capabilities.

Single country studies differentiated early on between internal and external (i.e. domestic and foreign) environmental/climate policy ambitions (e.g. Prittwitz 1984). For example, for much of the 1970s/1980s, the USA acted as an environmental leader on both the domestic and international levels (e.g. Rowlands 1995), while Japan's international ambitions did not match its progressive domestic environmental policy (e.g. Imura and Schreurs 2005). Importantly, CP state-centred and single country studies initially focused exclusively on highly developed liberal democracies. Studies did not initially view the transitional Central and Eastern European states, rapidly developing countries and developing countries as being capable of exhibiting environmental/climate leadership/pioneership, although scholars gradually are challenging this view (e.g. Torney 2015, Wurzel et al. 2017).

The *EU* seems to have provided particularly fertile ground for climate leadership/pioneership (e.g. Grubb and Gupta 2000, Jordan et al. 2010, Torney 2015, Wurzel et al. 2017, see also Jänicke and Wurzel 2019, – this volume). Schreurs and Tiberghien (2007) have argued that the EU's climate policy-making processes provide multi-level reinforcement mechanisms which can trigger the ratcheting upwards of the environmental standards set by the leaders or pioneers (see already Rehbinder and Stewart 1985).

From the early 1980s to the mid-1990s, the so-called Green Troika – Germany, the Netherlands and Denmark – acted as major driver for EU environmental policy development. The EU accession by Sweden, Finland and Austria in 1995 transformed the Green Trio into a green sextet and triggered a more systematic investigation of environmental leaders and pioneers and their impact (e.g. Héritier 1996, Andersen and Liefferink 1997, Liefferink and Andersen 1998, Börzel 2002, Wurzel 2008).

Within the climate leadership/pioneership literature the *subnational level* attracted scholarly attention only at a relatively late stage although there are exceptions (e.g. Freeman 1996, Bulkeley and Betsill 2005). One reason for this is the strong initial focus of the social science literature on international climate negotiations. As these negotiations became more protracted over time the scholarly search for evidence of climate leadership/pioneership shifted from the international level to national and subnational levels (e.g. Eckersley 2018, Kern 2019, Wurzel *et al.* 2019, – this volume).

Business is important for developing technological innovations and creating 'green' markets (e.g. Jänicke and Jacob 2002). However, businesses operating in jurisdictions with high regulatory standards may also have a more direct influence on policy by promoting the adoption of stringent standards at the international and/or supranational level for reasons of competitiveness (e.g. Vogel 1997, Liefferink and Wurzel 2018, Dupuis and Schweizer 2019). Moreover, business has frequently propagated the adoption of voluntary agreements and 'self-regulation' which theoretically fits polycentric approaches. However, business often 'remains the elephant in the room or perhaps just outside the door' (Grant 2011, p. 212) when it comes to the adoption of ambitious climate governance measures. We see this, for example, in the German automobile industry's resistance to more ambitious EU standards for carbon dioxide (CO_2) in the 2010s. As the literature remains ambivalent on whether business can effectively act as a climate leader or pioneer, this volume pays particular attention to the ability and/or willingness of business to offer climate leadership/pioneership (see Bach 2019, Biedenkopf *et al.* 2019, Dupuis and Schweizer 2019, Eikeland and Skjaerseth 2019).

Scholars widely recognise environmental *NGOs* (ENGOs) as important climate governance actors because they can help to raise public awareness, shape or even set the agenda and monitor adopted domestic policies and/or international treaties (e.g. Long *et al.* 2002, Bäckstrand *et al.* 2017). In contrast, *trade unions* are somewhat less prominent climate governance actors. According to Räthzel and Uzzell (2013, p. 5), 'new movements in the trade union movements' have shown 'a concern for nature by taking on climate change as an issue of trade union politics'. In particular unions in highly affluent countries have sought to combine environmental concerns with traditional trade union objectives through concepts such as 'climate justice' which are usually supported also by ENGOs.

Scientists (e.g. Rowlands 1995) and *epistemic communities* (e.g. Haas 1992) have played a more prominent role as climate leaders or pioneers. Although scientific insights are crucial for both recognising problems and solving them, Underdal (2000, p. 3) has argued that 'adequate knowledge about the problem itself and available response options is a *necessary* – although by no means a sufficient – condition for designing and operating *effective* international regimes'.

One should not underestimate the role of *individuals* – such as Al Gore or Laurent Fabius – in raising awareness, instigating climate governance action or competently chairing international negotiations (e.g. Parker and Karlsson 2014). Although some studies have flagged up the importance of individuals in international climate change politics (e.g. Rowlands 1995), the leadership/pioneership capabilities of individuals at different climate governance levels remains under-researched. There is, however, an extensive urban studies literature that identifies climate leadership/pioneership and entrepreneurship by individuals as being centrally important for local climate governance (e.g. Bulkeley and Betsill 2005, Jonas *et al.* 2011, Eckersley 2018, see also Kern 2019, Wurzel *et al.* 2019).

This volume is not able to assess whether *all* of the above mentioned types of actors have indeed been able to act as leaders and/or pioneers in climate governance. Instead it focuses on the EU (Jänicke and Wurzel 2019), corporate actors (Biedenkopf *et al.* 2019, Dupius and Schweizer 2019, Bach 2019, Eikeland and Skjærseth 2019) and cities (Kern 2019, Wurzel *et al.* 2019).

Why a leader/pioneer?

The question of *why* an actor tries to become an environmental leader or pioneer has received various answers. Much of the state-centred CP literature has focused on a wide variety of structural factors underlying the ambitions and motivations of environmental leaders and pioneers. It has viewed a high level of environmental problem pressure, high political salience of environmental issues and regulatory competition as important drivers for the emergence of environmental leaders and pioneers (e.g. Jänicke and Jacob 2002, Weidner *et al.* 2002, Liefferink *et al.* 2009). There is, however, often also a normative dimension, which we must consider when identifying and assessing environment leaders and pioneers. For example, while some states (e.g. France during its EU Presidency in 2008) have advocated on the EU level nuclear power to reduce greenhouse gas emissions, other states (e.g. Germany) have decided to phase out nuclear power in favour of renewable energy.

The environmental capacity literature has identified, among others, institutional, politico-administrative, informational-cognitive and technological capacities as core drivers of leadership/pioneership from states (e.g. Jänicke 2006). Other important explanatory factors include EU membership (e.g. Jänicke and Jacob 2002, Liefferink et al. 2009), corporatism (e.g. Crepaz 1995), high levels of economic affluence (e.g. Börzel 2002) as well as a wide range of issue and/or context specific factors (e.g. Jänicke and Weidner 1997, Liefferink et al. 2009).

Whether the same or similar factors which help to explain climate leadership/pioneership by states also apply to subnational and non-state actors (such as cities or business) is an unresolved question, which the contributions in this volume address. High environmental problem pressure, high political salience of environmental issues and regulatory competition are likely also to affect non-state actors. Creating and maintaining a green public image appears to be increasingly important for cities (e.g. Kern and Bulkeley 2009, Jonas et al. 2011, Eckersley 2018). The EU's annual European Green Capital Award and the European Energy Award are used as marketing tools by the holders (see Kern 2019, Wurzel et al. 2019). Important drivers for environmental innovations from business include pressure from environmentally concerned consumers and the creation of 'green' lead markets which can give 'green' companies a competitive advantage (e.g. Jänicke and Jacob 2002, see also Eikeland and Skjaerseth 2019, Dupuis and Schweizer 2019).

Another way of looking at why actors strive to become leaders or pioneers is to investigate the way their 'green' ambitions are structured. While drawing on the distinction between leaders and pioneers, we assess four possible combinations of an actor's *internal* and *external ambitions* (Prittwitz 1984, Liefferink et al. 2009, Liefferink and Wurzel 2017) on a scale which ranges from 'low' to 'high' (see Table 1). Although the analytical distinction between internal and external 'green' ambitions was originally developed only for states (Liefferink and Wurzel 2017), here we extend it also to non-state actors.

Table 1. Internal and external environmental ambitions of actors, leading to four ideal-typical positions.

External environmental ambitions	Internal environmental ambitions	
	Low	High
Low	(a) Laggard	(b) Pioneer
High	(c) Symbolic leader	(d) Substantive leader: - Constructive pusher - Conditional pusher

Source: Adapted from Liefferink and Wurzel (2017).

Table 1 distinguishes in ideal-typical fashion between the following four *positions*: low internal and low external ambitions lead actors to become *laggards* (or, at best, followers); the combination of high internal and low external ambitions turns actors into *pioneers* which try to 'go it alone' in particular if they consider themselves constrained by followers or laggards; low internal and high external ambitions turn actors into *symbolic leaders* which fail to back up their externally directed green ambitions with internal actions; the combination of high internal and high external ambitions turns actors into substantive *leaders* which seek others to adopt the same or at least similar ambitions. The analytical term *substantive leader* acts as an umbrella term that subsumes *constructive* and *conditional pushers*. While constructive pushers adopt unconditionally (i.e. unilaterally) ambitious internal environmental measures, conditional pushers take ambitious internal measures only if other actors adopt similar measures. The main reason for this is that conditional pushers are often concerned about the potentially negative economic impact of unilaterally adopted environmental standards. For a constructive pusher, environmental ambitions override economic concerns. The bottom row of Table 1 depicts the two possible cases of high external ambitions that can be associated with *leadership* although symbolic leaders fail to back up their high external ambitions with high internal ambitions, which stands in contrast to both sub-types of substantive leaders.

Pioneership/leadership refers either to actors who are *first* to introduce and/or propagate a certain policy measure or who exhibit the *highest* level of ambition. Being 'the first in class' and being 'the best in class' may not necessarily go together. In fact, followers, who may adopt even higher ambitions, can overtake the initial leader or pioneer (e.g. Burns 2003, Liefferink and Wurzel 2017).

How do leaders/pioneers act?

Leaders and pioneers may exert leadership in various ways. Building in particular on Young (1991), Liefferink and Wurzel (2017) and Wurzel et al. (2017), we distinguish the following four *types* of leadership[1]: structural, entrepreneurial, cognitive and exemplary. As we explain below, leaders can make use of all four types of leadership while pioneers are likely to offer primarily exemplary 'pioneership' by unintentionally setting a good example without having the explicit intention of attracting followers.

First, *structural* leadership has traditionally been associated with military power, especially by realist IR scholars (e.g. Waltz 1979). However, structural leadership has also been linked to economic power (e.g. Nye 2008). While few non-state actors (e.g. terrorist movements) possess military power, economic power can exist widely amongst non-state actors including in particular business (e.g. the capacity to create jobs) but also NGOs (especially if they have a large membership base) and consumers (e.g. purchasing power). Importantly, although structural leadership and power are closely related concepts, they are not identical (Young 1991, Nye 2008). Oberthür (2016, p. 83) has argued that '[p]ower and power structures have become an increasingly prominent consideration in analyses of international climate policy in the 21st century' while pointing out 'the rise of climate change to high politics, great power politics and even geopolitics'.

Burns's (1978, p. 19) argument that '[a]ll leaders are actual or potential power holders, but not all power holders are leaders' helps to explain why not all powerful states (e.g. the USA and China at various points in time) exhibit structural leadership in international climate governance. An actor possessing power becomes a structural leader only by mobilising its structural power in pursuit of collective goods. Power, in other words, is a necessary but not a sufficient condition for structural leadership. Morrison et al. (2017, p. 2) have criticised proponents of polycentricity for inadvertently rendering polycentrism as power-free because they ignore 'not only different types of power at play but also how their distribution may affect both governance processes and environmental outcomes'. Similarly, Singleton (2017, p. 1000) has argued that '[p]ower is a concept that remains largely underdeveloped within Ostrom's work rendering her themes "curiously apolitical" (Wall 2014, p. 480)'. By contrast, MLG often adopts a 'top-down view of subnational actors' (Fairbrass and Jordan 2004, p. 152) according to which supranational actors have greater decision-making powers despite the mutual dependencies which exists between them and subnational governance actors. In other words, MLG concepts assume that 'supranational actors play a decisive and proactive (i.e. entrepreneurial) rather than a subordinate role, in EU policy-making' (Fairbrass and Jordan 2004, p.151).

In addition to military and economic power, an actor's relative contribution to a particular environmental problem and/or its ability to offer solutions may also provide it with structural power. For instance, because China has been responsible for about 30% of the world's CO_2 emissions since the early 2010s it gained systemic relevance for and considerable structural power in international climate change politics. Similarly, business actors (e.g. the oil industry and wind energy sector) derive structural power from their relative contribution to the problem as well as their 'low-emission capacity', which Oberthür (2016, p. 85) defines as the 'ability to contribute to and benefit from the move to decarbonisation' (see also Bach 2019).

Secondly, *entrepreneurial* leadership involves the use of diplomatic and/or negotiating skills with a view to brokering compromises and agreements (Young 1991). An entrepreneurial leader is usually 'an agenda setter and popularizer who uses negotiating skill to devise attractive formulas and to broker interests' (Young 1991, p. 300). A wide range of state and non-state actors such as businesses and NGOs may employ entrepreneurial leadership. Polycentric concepts focus largely on site-specific conditions to assess 'the specific capabilities of individual actors and their potential to cooperate' (Dorsch and Flachsland 2017, p. 52). They tend to emphasise the importance of self-organisation in policy domain specific, decentralised decision-making system (Ostrom et al. 2012, Ostrom 2014). MLG concepts also reject the idea of a single or even central point of steering in terms of climate governance (Jordan et al. 2012, p. 52). Both polycentric and MLG concepts therefore refute the idea of monocentric governance. However, while MLG concepts emphasise the importance of 'baton passing' by environmental leaders/pioneers and 'multi-level reinforcement' by supranational actors for climate governance innovations to become sustainable (Schreurs and Tiberghien 2007), polycentric concepts argue that the emergence of self-organised cooperation as well as innovative experimentation and learning is most likely to emerge in decentralised, site-specific domains in which trust is high.

Thirdly, *cognitive* leadership involves defining or redefining ideas and concepts such as ecological modernisation, which postulates that ambitious environmental/climate measures may also benefit the economy, e.g. in the form of the 'green' or low carbon economy. Cognitive leadership may also relate to cause–effect relations and policy solutions through the provision of scientific and experiential knowledge regarding innovative climate measures (Haverland and Liefferink 2012). A wide range of actors (e.g. states, research institutes, business and NGOs) may be able to offer cognitive leadership. Such actors can form networks and epistemic communities (e.g. Haas 1992) that may enhance the cognitive leadership potential of actors. Both MLG and polycentric approaches consider 'the plurality of actors and levels and the complexity of their interactions not as obstacles but as an opportunity for innovation and interactive learning' (Jänicke 2017, p. 118, see also Marks and Hooghe 2004, p. 16, Ostrom et al. 2012, 2014). Ostrom (2014, p. 119) has advocated the adoption of 'a polycentric approach to the problem of climate change in order to gain the benefits at multiple scales as well as to encourage experimentation and learning from diverse policies adopted at multiple scales'. Similarly, Marks and Hooghe (2004, p. 16) argued that MLG structures 'facilitate innovation and experimentation'.

While powerful actors (e.g. large states and businesses) find it easier to acquire structural leadership capabilities than less powerful actors (e.g. small states and small cities), this is not necessarily the case for cognitive leadership capabilities. For example, some small EU member states (e.g. Denmark and the Netherlands) have been able to provide considerable cognitive leadership for EU environmental policy over a long period of time (e.g. Andersen and Liefferink 1997). From a cognitive leadership perspective it is therefore not surprising that the polycentricity as well as the urban governance literature have flagged up the importance of cities and regions as laboratories for experimentation and sources of innovation (e.g. Bulkeley and Betsill 2005, Ostrom et al. 2012, 2014). Subnational or regional actors have also been identified as important environmental innovators by the MLG literature (e.g. Fairbrass and Jordan 2004).

Fourthly, *exemplary leadership* (or leadership by example) refers to the intentional setting of examples for others while unintentional example setting is referred to as *exemplary pioneership* here and throughout this volume. Intentional exemplary leadership resembles what Grubb and Gupta (2000) have defined as *directional* leadership. *Intentional* exemplary leadership and directional leadership amount to a constructive pusher position (see Table 1). Constructive pushers intentionally put forward domestic policies as models for others. *Unintentional* example-setting, in contrast, refers to pioneers who do not seek to attract followers. State actors (e.g. national and subnational local governments) as well as non-state actors (e.g. business and NGOs, see Biedenkopf et al. 2019) may offer exemplary leadership/pioneership. These types of leadership/pioneership play an important role for both MLG and polycentric governance concepts, which pay particular attention to innovations by actors at 'lower' governance levels while arguing that they can be up-scaled to 'higher' governance levels (e.g. Schreurs and Tiberghien 2007, Ostrom et al. 2012).

Importantly, leaders can and usually do combine different leadership types (Young 1991, Parker and Karlsson 2014, p. 586, Wurzel et al. 2017). A leader can, for instance, simultaneously exert entrepreneurial leadership through coalition-building around a particular issue, cognitive leadership by supporting these efforts with scientific evidence, and exemplary leadership by acting as a model for others. The specific mix of different types of leadership employed by a particular actor, as well as the different ways in which they may interact varies across issues and may evolve over time.

The environmental/climate governance literature has used Burns's (1978, 2003) differentiation between transactional and transformational leadership (e.g. Liefferink and Wurzel 2017, Wurzel et al. 2017). Transactional leadership typically refers to low ambitions over a relatively short time horizon. In contrast, transformational leadership aims at profound or even 'revolutionary' changes. These are usually achievable only over a relatively long time horizon (cf. Burns 1978). However, 'revolutionary' leadership may also occur on a much shorter time scale. Germany's energy transition *(Energiewende)*, which demands the replacement of nuclear power by primarily renewable energy within a relatively short time period, is a fitting example. Furthermore, as Burns (1978, 2003) has pointed out, transactional leadership extending over an extremely long timescale may eventually trigger transformational change.

Differentiating between the degree of ambition and timescale allows for a more fine-grained analysis of leaders and pioneers. The contributions to this volume assess whether it is applicable not only to state but also to non-state actors. For example, Biedenkopf et al. (2019) and Dupuis and Schweizer (2019) show that highly innovative firms may have far-reaching *internal* ambitions in the short term while their *external* leadership ambitions are followed up over a much longer time scale.

Followers and followership

While the literature on environmental/climate leadership/pioneership reviewed above is extensive, scholars have paid much less attention to followers and followership, although notable exceptions exist (e.g. Torney 2014, 2015, for the general literature on leaders and followers see Rhodes and t'Hart 2014). This is perhaps understandable given the methodological and evidential challenges associated with convincingly identifying followers and followership. Much of the policy transfer and learning literature acknowledges that it is generally easier to identify actors who come up with policy innovations (i.e. act as their source) rather than to identify actors who emulate the leaders/pioneers and the mechanisms through which such emulation takes place (e.g. Tews et al. 2003). In order to convincingly identify followership, we must not only identify a purported leader/pioneer and an actor that has adopted similar policies or responses, but also prove that the leader/pioneer and follower are causally linked.

As Rhodes and t'Hart (2014, p. 6) have pointed out '[t]here is now a growing body of thought and research that understands leadership as an interactive process between leaders and followers'. Followership is crucial in MLG and polycentric climate governance. However, polycentric governance approaches generally assume relatively high levels of autonomy of actors within and between polycentric governance systems (see above). At first sight, leader–follower relationships seem much more likely to emerge *within* rather than *between* site-specific polycentric governance domains. However, as argued above, polycentric governance actors may still function within an overarching MLG context (Liefferink and Wurzel 2018). Furthermore, even if actors within polycentric governance systems enjoy considerable levels of formal autonomy, their relationships with other actors may nonetheless be characterised by interconnectedness, providing opportunities for the emergence of leader–follower dynamics. Morrison et al. (2017, p. 2) have defined a polycentric system as 'made up of many autonomous units that are formally independent of one another but which choose to act in ways that take account of others through self-organized processes of cooperation and conflict resolution'. Nevertheless, pioneers, who do not intentionally set out to gain followers, should be more prevalent than leaders in polycentric governance systems, which are characterised by relatively autonomous market-like or self-organising decision-making structures (Liefferink and Wurzel 2018). The policy transfer, diffusion and learning literature covers well why and how pioneers may unintentionally and leaders intentionally attract followers (e.g. Tews *et al.* 2003, see also Jänicke and Wurzel 2019).

An attempt to conceptualise followership poses a number of questions (see Torney 2019). First, *who* follows? In principle, followers can emerge in response to leaders in all of the categories of actors set out above: states, supranational and subnational actors, businesses, NGOs, trade unions, scientists and individuals. In order to count as followership, the follower must in some meaningful way adopt the same or a substantively similar approach to a particular climate/environmental problem. This limits the potential for leader–follower dynamics that cut across actor categories, and makes within-actor category leader–follower dynamics (e.g. state–state, business–business, NGO–NGO) more likely. We could consider the success of a social enterprise such as FairPhone in providing exemplary leadership for other corporations as followership on the part of those other corporations, though, as Biedenkopf *et al.* show in their contribution to this volume, structural constraints limit the possibility of such followership materialising. Wurzel *et al.* (2019), who assess whether structurally disadvantaged cities can act as leaders/pioneers or merely as followers of more affluent leader cities, argue that even for the same type of actor (i.e. cities) further differentiation may be necessary to establish the dynamics of leader/pioneer–followership relations. Only contextualised empirical research can accomplish this differentiation.

Various theoretical approaches provide different perspectives on the potential for leader–follower relationships across actor categories. For state-centric approaches, normally only state actors (i.e. 'within-category' actors) are able to enter into analytically meaningful leader–follower relations. By contrast, polycentric governance approaches allow for leader–follower relations between potentially all actor categories (e.g. individuals, businesses, NGOs and state actors). Meanwhile, MLG concepts focus on leader–follower relations at different governance levels and usually emphasise the role of public actors such as state, supranational and subnational actors. MLG concepts therefore arguably sit somewhere in between polycentric and state-centric concepts as regards analytically meaningful leadership–followership relations. However, as discussed above, some variants of MLG concepts are quite close to polycentric concepts concerning the importance they attach to non-state actors as governance actors.

A second question concerns the pathways through which the leader–follower dynamic plays out. Followership can emerge in a variety of ways and for different reasons. It is helpful to distinguish between followership that springs from a logic of consequences and a logic of appropriateness (March and Olsen 1998). This also, to some extent, maps on to the types of leadership identified above. Drawing on a logic of consequences, followers may be induced to follow structural leadership through material incentives that alter their cost-benefit calculations. From a logic of appropriateness perspective, by contrast, followership can materialise through learning and emulation, including new ideas promulgated by cognitive leaders or from exemplary leaders perceived to provide innovative solutions. In such cases, followers follow leaders not because they are incentivised to do so, but because they believe the models a leader provides to be superior in some way and worthy of followership. Followership may emerge in response to entrepreneurial leadership on the basis of both a logic of consequences and a logic of appropriateness.

A third question concerns the conditions under which leaders attract followers. We can identify relevant factors with respect to both the leader and the purported follower. Again, it is helpful to distinguish between a logic of consequences and a logic of appropriateness. In the case of the leader–follower dynamic with respect to structural leadership (logic of consequences), the source of structural power matters, as does the degree of power asymmetry between the leader and follower (see Morrison et al. 2017). Economic power due to strong market asymmetries may give other actors little choice but to follow (although structural leadership usually does not amount to outright coercion which would leave no options to other actors but to follow the leader), whereas less asymmetric and/or symmetric market-like relations are likely to give purported followers considerably more leeway to pursue alternative options. In the case of a logic of appropriateness, the perceived legitimacy of an exemplary or cognitive leader is likely to be critical. Does the leader follow up its external ambitions by internal actions, or do other actors perceive its leadership to be merely symbolic, and are the models and knowledge that the purported leader provides viewed as authoritative (Parker and Karlsson 2010)? The extent to which the idea/approach that the leader promulgates resonates with pre-existing domestic norms and beliefs of the follower, and the extent to which the leader has sought, by way of cognitive leadership, to build congruence between new and pre-existing norms and beliefs are also likely to be important (e.g. Checkel 2005).

The questions of *who* follows, through what *pathways*, and *under what circumstances* are followers likely to follow, provide a roadmap for developing systematic research on the under-researched leader/pioneer–follower relationship. Pursuing answers to these questions allows us to fill a gap in the literature.

Conclusion

States are no longer, if they ever were, seen as the only actors capable of acting as climate leaders/pioneers (e.g. Young 1991, Underdal 1994, Wurzel *et al.* 2017). Scholars have also identified a wide range of subnational and non-state actors (e.g. cities and businesses) as putative agents of change. Importantly, climate leaders or pioneers often must act (either simultaneously or sequentially) at different levels of climate governance to be able to achieve their internal and/or external ambitions and, in the case of leaders, to attract followers.

State-centred, MLG and polycentric governance concepts place different emphasis on the roles played by different types of leadership/pioneership in climate governance. Although we should not exaggerate the differences between these three conceptual perspectives, it seems clear that state-centred theories (especially in the IR literature) attach proportionately greater emphasis to state actors and inter-state relations. Without downplaying the role of cognitive, entrepreneurial and exemplary leadership in such relations, a state-centric perspective is likely to pay particular attention to structural leadership. MLG concepts emphasise the importance of supranational EU and sub-state actors and their network-like relationships (which stretch across different levels of governance and require in particular entrepreneurial leadership) although MLG also pays attention to structural, cognitive and exemplary leadership. Polycentric governance approaches argue that successful climate governance depends largely on the existence of strong relatively autonomous decentralised and/or local decision-making centres which allow for societal self-coordination or even self-governing under conditions of trust. Polycentric settings thus encourage experimentation and learning-by-doing, i.e. cognitive and exemplary leadership, which, if successful, can be scaled up in a bottom-up fashion to other actors or higher levels of climate governance. Polycentric governance approaches arguably pay least attention to structural leadership.

Why actors develop into climate leaders/pioneers depends on their internal and external ambitions as well as on a set of more structural drivers including problem pressure, the political/public salience of climate change and competitive pressures which need to be established empirically. In this introduction, we have differentiated between structural, entrepreneurial, cognitive and exemplary leadership/pioneership which are assessed in the contributions that follow in this volume. The analytical differentiation into different types of leadership/pioneership helps to explain why some actors which have relatively little structural power may nevertheless become relatively influential climate governance actors capable of showing leadership or pioneership. The reason for this is that such actors may be able to compensate at least partly for their lack of structural leadership capacity with entrepreneurial and/or cognitive leadership/pioneership.

Whether leaders attract followers has, to a certain extent, been the 'poor relation' in the literature. While polycentric governance concepts focus strongly on 'between category' leader–follower relations, state-centric and MLG concepts assume that followers primarily emerge in response to leadership by actors from the same actor category. In other words, within-category leader–follower relationships are most likely while between-category 'leader-follower' like relations are conceptually best termed as *influence*. Within category followership can emerge as a result of power asymmetries (e.g. market asymmetries) or the provision of incentives (logic of consequences), deriving from structural leadership, on the one hand, or the power of attraction (logic of appropriateness), deriving primarily from cognitive and exemplary leadership, on the other.

The other contributions in this volume focus in particular on the EU (Jänicke and Wurzel), corporate actors (Biedenkopf *et al.* 2019, Dupius and Schweizer 2019, Bach 2019, Eikeland and Skjærseth 2019) and cities (Kern 2019, Wurzel *et al.* 2019). While all contributions assess leadership and pioneership within MLG and/or polycentric governance systems, Torney's contribution (2019) focuses in particular on followers and followership.

Note

1. Many of the early IR leadership studies put forward a threefold typology. For example, Young (1991) drew on structural, intellectual and entrepreneurial leadership while Underdal (1994) referred to coercive, unilateral and instrumental leadership. Malnes (1995) added directional leadership, which amounts to leadership by example, as a fourth analytical category (see also Grubb and Gupta 2000, Parker and Karlsson 2010, 2014).

Acknowledgments

The authors are grateful to the Innovations in Climate Governance (INOGOV) programme which funded a workshop on 'Pioneers and Leaders in Polycentric Climate Governance (PiLePoC) in Hull on 15–16 September 2016 where early versions of contributions to this volume were presented. Rudi Wurzel thanks the British Academy (grant no. SG 131240) and the University of Hull for additional funding. Diarmuid Torney thanks Dublin City University for support from the Faculty of Humanities and Social Sciences Journal Publication Scheme. We are very grateful to Louise FitzGerald for extensive editorial assistance and to the referees as well as to the journal editors, Anthony Zito and Chris Rootes, for their very helpful comments. The usual disclaimer applies.

Disclosure statement

No potential conflict of interest was reported by the authors.

Funding

The authors are grateful to the EU COST funded Innovations in Climate Governance (INOGOV) programme which funded a workshop on 'Pioneers and Leaders in Polycentric Climate Governance (PiLePoC) in Hull on 15–16 September 2016 where early versions of contributions to this volume were presented. Rudi Wurzel thanks the British Academy (grant no. SG 131240) and the University of Hull for additional funding. Diarmuid Torney thanks Dublin City University for financial support.

ORCID

Rüdiger K.W. Wurzel http://orcid.org/0000-0001-5873-4232
Duncan Liefferink http://orcid.org/0000-0002-3594-3274
Diarmuid Torney http://orcid.org/0000-0003-4156-9044

References

Andersen, M.S. and Liefferink, J.D., eds., 1997. *European environmental policy: the pioneers*. Manchester: Manchester University Press.
Bach, M., 2019. The oil and gas sector: from climate laggard to climate leader? *Environmental Politics*, 28 (1).
Bäckstrand, K., *et al.*, 2017. Non-state actors in the new landscape of international cooperation. *Environmental Politics*, 26 (4), 561–799. doi:10.1080/09644016.2017.1327485
Biedenkopf, K., Bachus, K., and van Eynde, S., 2019. Environmental, climate and social leadership of small enterprises: FairPhone's step-by-step approach. *Environmental Politics*, 28 (1).
Börzel, T.A., 2002. Pace-setting, foot-dragging and fence-sitting. *Journal of Common Market Studies*, 40 (2), 193–214.
Bulkeley, H. and Betsill, M., 2005. Rethinking sustainable cities: multilevel governance and the 'urban' politics of climate change. *Environmental Politics*, 14 (1), 42–63. doi:10.1080/0964401042000310178
Burns, J.M., 1978. *Leadership*. New York: Harper & Row.
Burns, J.M., 2003. *Transforming leadership*. New York: Grove Press.
Checkel, J.T., 2005. International institutions and socialization in Europe: introduction and framework'. *International Organization*, 59 (4), 801–826. doi:10.1017/S0020818305050289
Crepaz, M., 1995. Explaining national variations of air pollution levels. *Environmental Politics*, 4 (3), 391–414. doi:10.1080/09644019508414213
Dorsch, M.J. and Flachsland, C., 2017. A polycentric approach to global climate governance. *Global Environmental Politics*, 17 (2), 45–64. doi:10.1162/GLEP_a_00400

Dupuis, J. and Schweizer, R., 2019. Climate pushers or symbolic leaders? The limits to corporate climate leadership by food retailers. *Environmental Politics*, 28 (1).

Eckersley, P., 2018. Who shapes local climate policy? Unpicking governance arrangements in English and German cities. *Environmental Politics*, 27 (1), 139–160. doi:10.1080/09644016.2017.1380963

Eikeland, P.O. and Skjærseth, J.B., 2019. Oil and power industries' responses to EU emissions trading: laggards or low-carbon leaders?. *Environmental Politics*, 28 (1).

Fairbrass, J. and Jordan, A., 2004. Multi-level governance and environmental policy. *In*: I. Bache and M. Flinders, eds. *Multi-level governance*. Oxford: Oxford University Press, 147–164.

Freeman, C., 1996. Local government and emerging models of participation in the Local Agenda 21 process. *Journal of Environmental Planning and Management*, 39 (1), 65–78. doi:10.1080/09640569612679

Grant, W., 2011. Business: the elephant in the room. *In*: R.K.W. Wurzel and J. Connelly, eds. *The European Union as a leader in international climate change politics*. London: Routledge, 197–213.

Grubb, M. and Gupta, J., 2000. Climate change, leadership and the EU. *In*: J. Gupta and M. Grubb, eds. *Climate change and European leadership*. Dordrecht: Kluwer, 3–14.

Haas, P.M., 1992. Introduction: epistemic communities and international policy coordination. *International Organization*, 46 (1), 1–35. doi:10.1017/S0020818300001442

Haverland, M. and Liefferink, D., 2012. Member state interest articulation in the commission phase. *Journal of European Public Policy*, 19 (2), 179–197. doi:10.1080/13501763.2011.609716

Héritier, A., 1996. The accommodation of diversity in European policy-making and its outcomes: regulatory policy as a patchwork. *Journal of European Public Policy*, 3 (2), 149–167. doi:10.1080/13501769608407026

Homsy, G.C. and Warner, M.E., 2014. Cities and sustainability: polycentric action and multilevel governance. *Urban Affairs Review*, 49 (1), 1–28.

Hooghe, L., ed., 1996. *Multi-level governance and European integration*. Oxford: Clarendon Press.

Imura, H. and Schreurs, M.A., eds., 2005. *Environmental policy in Japan*. Cheltenham: Edward Elgar.

Jänicke, M., 2006. Trend setters in environmental policy: the character and role of pioneer countries. *In*: M. Jänicke and K. Jacob, eds. *Environmental governance in global perspective*. Berlin: Environmental Policy Research Centre, 51–66.

Jänicke, M., 2017. The multi-level system of global climate governance – the model and its current state. *Environmental Policy and Governance*, 27, 108–121. doi:10.1002/eet.1747

Jänicke, M. and Jacob, K. 2002. *Ecological modernisation and the creation of lead markets*. FFU Report 02-03, Berlin: Environmental Policy Research Centre. doi:10.1044/1059-0889(2002/er01)

Jänicke, M. and Weidner, H., 1997. *National environmental policies*. Berlin: Springer.

Jänicke, M. and Wurzel, R., 2019. Leadership and lesson-drawing in the European Union's multilevel climate governance system. *Environmental Politics*, 28 (1).

Jonas, A.E.G., Gibbs, D., and While, A., 2011. The new urban politics as a politics of carbon control. *Urban Studies*, 48, 2537–2544.

Jordan, A., et al., 2012. Understanding the paradoxes of multilevel governing: climate change policy in the European Union. *Global Environmental Change*, 12 (2), 43–66.

Jordan, A., et al., eds., 2018. *Governing climate change: polycentricity in action*. Cambridge: Cambridge University Press.

Jordan, A., et al., eds., 2010. *Climate change policy in the European Union*. Cambridge: Cambridge University Press.

Kern, K., 2019. Cities as leaders in EU multi-level climate governance? Embedded upscaling of local experiments in Europe. *Environmental Politics*, 28 (1).

Kern, K. and Bulkeley, H., 2009. Cities, Europeanization and multi-level governance: governing climate change through transnational municipal networks. *Journal of Common Market Studies*, 47 (2), 309–332. doi:10.1111/j.1468-5965.2009.00806.x

Liefferink, D. and Andersen, M.S., 1998. Strategies of the "green" member states in EU environmental policy-making. *Journal of European Public Policy*, 5 (2), 254–270. doi:10.1080/135017698343974

Liefferink, D., et al., 2009. Leaders and laggards in environmental policy. *Journal of European Public Policy*, 16 (5), 677–700. doi:10.1080/13501760902983283

Liefferink, D. and Wurzel, R.K.W., 2017. Environmental leaders and pioneers: agents of change? *Journal of European Public Policy*, 24 (7), 651–668. doi:10.1080/13501763.2016.1161657

Liefferink, D., et al., 2018. Leaders and pioneers in polycentric governance. In: A. Jordan, ed. *Governing climate change: polycentricity in action*. Cambridge: Cambridge University Press, 135–151.

Long, T., Slater, L., and Singer, S., 2002. WWF: European and global climate policy. In: R. Pedler, ed. *European Union lobbying*. Basingstoke: Palgrave, 87–103.

Malnes, R., 1995. Leader' and 'entrepreneur' in international negotiations: A conceptual analysis. *European Journal of International Relations*, 1 (1), 87–112. doi:10.1177/1354066195001001005

March, J.G. and Olsen, J.P., 1998. The institutional dynamics of international political orders. *International Organization*, 52 (4), 943–969. doi:10.1162/002081898550699

Marks, G., 1993. Structural policy and multi-level governance in the EC. In: A. Cafruny and G. Rosenthal, eds. *The state of the European community*. Boulder: Lynne Rienner, 391–411.

Marks, G. and Hooghe, L., 2004. Contrasting visions of multi-level governance. In: I. Bache and M. Flinders, eds. *Multi-level governance*. Oxford: Oxford University Press, 15–30.

Morrison, T.H., et al., 2017. Mitigation and adaptation in polycentric systems: sources of power in the pursuit of collective goals. *WIREs Climate Change*, 7, 1–16. doi:10.1002/wcc.479

Nye, J., 2008. *The powers to lead*. Oxford: Oxford University Press.

Oberthür, S., 2016. Reflections on global climate politics post Paris: power, interests and polycentricity. *The International Spectator*, 51 (4), 80–94. doi:10.1080/03932729.2016.1242256

Ostrom, E., et al., 2012. The future of the commons: beyond market failure and government regulation. In: E. Ostrom, ed. *The Future of the Commons*. London: The Institute of Economic Affairs, 68–83.

Ostrom, E., 2014. A polycentric approach for coping with climate change. *Annals of Economics and Finance*, 15 (1), 97–134.

Parker, C.F. and Karlsson, C., 2010. Climate change and the European Union's leadership moment: an inconvenient truth? *Journal of Common Market Studies*, 48 (4), 923–943. doi:10.1111/j.1468-5965.2010.02080.x

Parker, C.F. and Karlsson, C., 2014. Leadership and international cooperation. *In*: R.A.W. Rhodes and P. 'T Hart, eds. *The handbook of political leadership*. Oxford: Oxford University Press, 580–594.

Prittwitz, V.V., 1984. *Umweltaußenpolitik*. Frankfurt: Campus Verlag.

Räthzel, N. and Uzzell, D., 2013. Mending the breach between labour and nature: a case for environmental labour studies. *In*: N. Räthzel and D. Uzzell, eds. *Trade unions in the green economy*. London: Earthscan, 1–11.

Rehbinder, E. and Stewart, R., 1985. *Integration through law*. Berlin: Walter de Gruyter.

Rhodes, R.A.W. and 'T Hart, P., 2014. Puzzles of political leadership. *In*: R.A.W. Rhodes and P. 'T Hart, eds. *The handbook of political leadership*. Oxford: Oxford University Press, 1–21.

Rowlands, I., 1995. *The politics of global atmospheric change*. Manchester: Manchester University Press.

Schreurs, M.A. and Tiberghien, Y., 2007. Multi-level reinforcement: explaining European Union leadership in climate change mitigation. *Global Environmental Politics*, 7 (4), 19–46. doi:10.1162/glep.2007.7.4.19

Singleton, B.E., 2017. What is missing from Ostrom? Combining design principles with the theory of sociocultural viability. *Environmental Politics*, 26 (6), 994–1014. doi:10.1080/09644016.2017.1364150

Stephenson, P., 2013. Twenty years of multi-level governance. *Journal of European Public Policy*, 20 (6), 817–837. doi:10.1080/13501763.2013.781818

Tews, K., Busch, P., and Jörgens, H., 2003. The diffusion of new environmental policy instruments. *European Journal of Political Research*, 42 (4), 569–600. doi:10.1111/1475-6765.00096

Torney, D., 2014. External perceptions and EU foreign policy effectiveness: the case of climate change. *Journal of Common Market Studies*, 52 (6), 1358–1373. doi:10.1111/jcms.12150

Torney, D., 2015. *European climate leadership in question*. Cambridge. Massachusetts: MIT Press.

Torney, D., 2019. Follow the leader? Conceptualising the relationship between leaders and followers in polycentric climate governance. *Environmental Politics*, 28 (1).

Underdal, A., 1994. Leadership theory: rediscovering the arts of management. *In*: W.I. Zartman, ed. *International multilateral negotiation*. San Francisco: Jossey-Bass, 178–197.

Underdal, A., 2000. Science and politics. *In*: S. Andresen, T. Skodvin, A. Underdal, and J. Wettestad, eds. *Science and politics in international environmental regimes*. Manchester: Manchester University Press, 1–21.

Vogel, D., 1997. *Trading up. Consumer and environmental regulation in a global economy*. Cambridge M.A.: Harvard University Press.

Wall, D., 2014. *The sustainable economics of Elinor Ostrom*. London: Routledge.

Waltz, K., 1979. *Theory of international politics*. Reading: Addison-Wesley.

Weidner, H., Jänicke, M., and Jörgens, H., eds., 2002. *Capacity building in national environmental policy*. Berlin: Springer.

Wurzel, R., *et al.*, 2019. Climate pioneership and leadership in structurally disadvantaged maritime port cities. *Environmental Politics*, 28 (1).

Wurzel, R.K.W., 2008. Environmental policy: EU actors, leader and laggard states. *In*: J. Hayward, ed. *Leaderless Europe*. Oxford: Oxford University Press, 66–88.
Wurzel, R.K.W., Connelly, J., and Liefferink, D., eds., 2017. *The European Union in international climate change politics*. London: Routledge.
Young, O.R., 1991. Political leadership and regime formation: on the development of institutions in international society. *International Organization*, 45 (3), 281–308. doi:10.1017/S0020818300033117

Leadership and lesson-drawing in the European Union's multilevel climate governance system

Martin Jänicke and Rüdiger K.W. Wurzel

ABSTRACT
The important role that climate leaders and leadership play at different levels of the European Union (EU) multilevel governance system is exemplified. Initially, climate leader states set the pace with ambitious policy measures that were adopted largely on an ad hoc basis. Since the mid-1980s, the EU has developed a multilevel climate governance system that has facilitated leadership and lesson-drawing at all governance levels including the local level. The EU has become a global climate policy leader by example although it had been set up as a 'leaderless Europe'. The resulting 'leadership without leader' paradox cannot be sufficiently explained merely by reference to top-level EU climate policies. Local-level climate innovations and lesson-drawing have increasingly been encouraged by the EU's multilevel climate governance system which has become more polycentric. The recognition of economic co-benefits of climate policy measures has helped to further the EU's climate leadership role.

Introduction

Leadership and lesson-drawing by followers has a long history in environmental policy.[1] It has become particularly important for European Union (EU) climate policy (e.g. Jänicke 2005, 2017b, Oberthür and Kelly 2008, Wurzel et al. 2017). Environmental leaders are actors such as national governments that are first in finding solutions for environmental problems (Andersen and Liefferink 1997). If leaders attract followers due to lesson-drawing (Rose 1993), then they become *leaders by example* (Liefferink and Wurzel 2017, Wurzel et al. 2017).[2] According to Rose (1993) *lesson-drawing* takes place when an effective policy solution is transferred from one place to another. Therefore, *lesson-drawing* requires followers who emulate an innovative solution (or at least significant elements of it) used elsewhere.

Lesson-drawing may offer followers a shortcut to innovative solutions and/ or reduce their domestic 'learning costs'.

The academic literature has identified additional factors which may act as drivers explaining why states or substate actors adopt the same or similar policies, programmes and instruments (e.g. Jordan *et al.* 2003, 2013). First, policy *convergence* occurs when similar states or substate actors adopt the same or similar policy solution independently from each other. This is most likely to occur when similar types of actors face the same or similar problems. Second, transnational networks, which can be widely found within the EU, may facilitate the transfer of environmental innovation. Third, cooperation and/or competition between states or substate actors can lead to the adoption of similar innovations. Radaelli (2000, p. 26) has called the EU's competitive single European market (SEM) a 'massive transfer platform' for shifting policies, programmes or instruments between member states.

Largely due to space constraints, we focus primarily on *lesson-drawing* from climate leaders (rather than also on policy convergence and/or regulatory competition) which, we argue, is of central importance for EU climate governance. We try to identify and explain cases of best practice within the EU multilevel climate governance system that has developed increasingly more advanced opportunity structures for lesson-drawing at different climate governance levels. We distinguish between the following four *types* of leadership: *structural* leadership which relates mainly to economic power; *entrepreneurial* leadership which relies heavily on diplomatic, negotiating and bargaining skills; *cognitive* leadership which depends primarily on knowledge and expertise; and *exemplary* leadership which occurs when actors intentionally or unintentionally set an example for others (cf. Liefferink and Wurzel 2017, Wurzel *et al.* 2017). We further assess whether EU climate governance exhibits mainly a *transformative* or a *transactional* (i.e. incremental) leadership *style* (Liefferink and Wurzel 2017, see also Wurzel *et al.* 2019, this volume).

Although Hayward (2008) has characterised the EU as a 'leaderless Europe', it has frequently offered exemplary global climate leadership (Schreurs and Tiberghien 2007, Oberthür and Kelly 2008, Jordan *et al.* 2012, Wurzel *et al.* 2017). The resulting 'leaderless leader' paradox in climate governance, therefore, needs explaining. We argue that merely focusing on top-level governance decisions and legally binding laws, which have a *direct* effect on member states, cannot explain sufficiently climate governance innovations within the EU's multilevel climate governance system. Instead, *indirect* effects may also play an important role and help to explain why the EU's overall climate governance performance is often better than what the top level of the EU climate governance system has decided (e.g. Schreurs and Tiberghien 2007).

The role of national leaders and early followers

National leaders have played an influential role even before the EU adopted a common environmental policy in the early 1970s (e.g. Rehbinder and Stewart 1985) and a common climate policy in the early 1990s (e.g. Jordan *et al.* 2012). In the late 2010s, national climate leaders still acted as major drivers of the EU's climate leadership (Oberthür and Kelly 2008, Wurzel *et al.* 2017). However, as the EU multilevel climate governance system has matured over the years, it has arguably become more 'polycentric' (Ostrom 2010, 2014). The growing polycentric features of the EU's governance system have provided the subnational level with new roles and functions in climate governance innovation (CoR 2014, Ostrom 2014). While there is a well-established literature on the Europeanisation of member states' environmental policies (e.g. Hèritier *et al.* 1996, Jordan and Liefferink 2004), scholars have paid much less attention to the less tangible impact of the EU's multilevel climate governance system on cities and regions, although there are important exceptions (e.g. Kern and Bulkeley 2009, Bendlin *et al.* 2016, see also Kern 2019, Wurzel *et al.* 2019, both this volume).

In the early 1970s, Sweden and the United States acted as early environmental leaders with Japan and Germany as the main early followers. Environmental leaders were able to use events such as mass demonstrations against air pollution in the United States, massive public pressure caused by environmental lawsuits in Japan (e.g. minamata, itai-itai and yokkaichi asthma) as windows of opportunity for (environmental) policy change as well as changes in government which occurred, for example, when a reform-minded Social Democratic-Liberal (Social Democratic Party (SPD) – Free Democratic Party (FDP)) coalition came into government in Germany in 1969 (Jänicke and Weidner 1997). Sweden, the United States, Japan and Germany as well as Denmark also introduced the largest number of environmental policy innovations (e.g. new institutions and laws) between 1970 and 1985 (Jänicke 2005). Initially, there was a strong international demonstration effect by the United States regarding new institutions and laws. For example, other states examined closely the Environmental Protection Agency (EPA) and early US air and water pollution laws, adopting similar laws at a later stage (Wurzel 2002, pp. 244–5).

National climate leaders

In the late 1980s, global *climate* governance started with initiatives from national leaders whose policy innovations greatly facilitated the adoption of the United Nations Framework Convention on Climate Change (UNFCCC) at the 1992 UN Rio conference. National European climate policy leaders, which we will assess briefly in this section, have been selected according to the ambition of their Kyoto Protocol targets for 2008/2012, the ambition of their targets for 2020/2025, their greenhouse gas emission (GHGE) reductions between 1990 and 2015 and the persistence of their leadership over a long period of time.

Denmark, Germany, Sweden and the United Kingdom fulfil all four criteria. The climate policies of these four countries were conceived already in the late 1980s/early 1990s. Their Kyoto Protocol targets for 1990–2008/2012 were the most ambitious with the exception of Sweden, which had already undertaken early actions to cut GHGE. Under the EU's so-called burden sharing agreement, which divided up the EU's GHGE reduction target (−8%) into differentiated national targets, both Germany and Denmark accepted reduction targets of −21% and the United Kingdom −12.5% (e.g. Wurzel et al. 2017).

These four member states also adopted relatively ambitious long-term national GHGE reductions targets. The United Kingdom set itself a national GHGE reduction target of −50% by 2025 while Denmark, Germany and Sweden each accepted reduction targets of −40% by 2020. The climate policies of these four countries have long been exceptional, and their GHGE reductions were the most ambitious of all Western European countries between 1990 and 2015. The United Kingdom and Germany alone accounted for 47.9% of the EU's total net decrease in GHGE between 1990 and 2015 (EEA 2017). The Netherlands was also an early leader, which already adopted a climate policy chapter in its influential 1989 National Environmental Policy Plan (NEPP). In 1989, the Netherlands introduced feed-in tariffs before Germany (1990), and Denmark (1993) followed the Dutch lead (Jacobs 2012). Due to space constraints, we focus here on Denmark, Germany and the United Kingdom.

Dimensions of climate leadership

Germany

Germany developed into a climate leader already in the mid-1980s, since when it has persistently acted as a climate leader, although Germany has struggled to comply with its ambitious 2020 GHGE reduction target (see below). Germany provided all main leadership types identified by Liefferink and Wurzel (2017) – structural, entrepreneurial, cognitive and exemplary (Jänicke 2017b). In 1986, the (West) German government started to adopt its first climate policy measures following an initiative by the federal Upper House (*Bundesrat*). In 1987 – a year of federal elections that resulted in increased votes for the Green Party – the national parliament (*Bundestag*) set up an Enquete Commission on Preventive Measures to Protect the Earth's Atmosphere, while the government adopted a CO_2 Reduction

Programme with cross-party support in 1990. The 1990 Commission report offered a broad overview of the findings from climate change research while proposing ambitious GHGE reduction targets not only for Germany but also for the EU (Deutscher Bundestag 1990). The report thus offered cognitive climate leadership while demanding exemplary leadership from the German government and the EU.

The first Conference of the Parties (COP1) to the UNFCCC took place in Berlin in 1995. The German government under Chancellor Helmut Kohl (Christian Democratic Union (CDU)) offered significant entrepreneurial leadership while presenting an ambitious German GHGE reduction target of −25% by 2005, thus also offering exemplary leadership. Germany was an active player at all COPs which followed, while also pushing climate issues at G7 and G20 meetings and on the EU level especially when holding the rotating presidency in these international settings (Wurzel 2010). Further examples of exemplary leadership include the rapid uptake of 'clean power' in the form of renewable energy, significant CO_2 emission reductions and economically successful climate policies. Importantly, cooperatives and local communities have played a strong role for many German climate innovations. Germany developed a lead market for wind and photovoltaic (PV). This matters in terms of structural leadership because Germany, as the largest economy in the EU, was thus able to exert competitive pressure within the SEM and on the global market for clean energy technologies, although with less success in recent years. By 2013, Germany had 17% of the global clean energy market (Jänicke 2017a). However, like all environmental/climate leaders, Germany also has its blind spots, as we can see, for example, from its continued reliance on coal-fired power stations and the German automobile industry's relatively poor fuel efficiency record. This was the main reason for the German government's opposition to the EU Commission's 2014 proposal for more ambitious CO_2 emission standards.

The United Kingdom

The United Kingdom has adopted a leadership role in climate policy since the early 1990s (Rayner and Jordan 2017). Prime Minister Margaret Thatcher's decision to drastically reduce energy generation from coal-fired power stations while expanding the use of gas occurred primarily for cost and political reasons (to curb the influence of the miners' union) rather than environmental reasons. As Rayner and Jordan (2017, p. 175) have pointed out, 'the ensuing "dash for gas" had the completely unintended effect of lowering the UK's emissions throughout the 1990s', paving the way for the United Kingdom's climate leadership. In 1990, the United Kingdom introduced the Non-Fossil Fuel Obligation. The Fuel Duty Escalator (1993) and the Climate Change Levy (1999) followed. In 2002, the United Kingdom exhibited exemplary leadership by adopting a national emission trading scheme (ETS) in order to gain early practical experience and influence the rules of the EU ETS, which became operational in 2005. The United Kingdom has also been a local-level climate leader as a large number of its cities have adopted climate change mitigation and/or adaptation plans (Kern and Bulkeley 2009).

The United Kingdom showed exemplary leadership by adopting the most ambitious non-binding, national long-term GHGE reduction targets of all EU member states. Between 1990 and 2015, the United Kingdom had already achieved a 36.6% reduction of GHGE (EEA 2017), and in 2008, it introduced the world's first Climate Change Act, which stipulated a binding GHGE reduction target of −80% by 2050.

The United Kingdom's 2002 Energy Efficiency Commitment was an important innovation in Europe. Under successive Labour governments (1997–2010), climate policy was of central importance. Especially after 2005, Prime Minister Tony Blair conceived climate policy as a *business opportunity*, which could turn the United Kingdom into a successful exporter of low-carbon technologies. Under a Conservative-Liberal Democrat coalition government (2010–2015), there was initially a strong degree of continuity in terms of the United Kingdom's EU and global climate leadership, but this has since come 'under threat' (Rayner and Jordan 2017, p. 177). For example, the fuel duty escalator was scrapped in 2011. The United Kingdom's decision to leave the EU – Brexit – in 2019 has created further uncertainty.

Denmark

Denmark has been called the 'motherland' of the clean energy transformation (Meyer and Koefoed 2003). The Danes have already introduced regular energy plans supporting renewable energy and energy efficiency since 1976. Denmark adopted a CO_2 tax in 1992 at time when the United Kingdom vetoed the European Commission's proposal for an EU-wide CO_2/energy tax on sovereignty grounds. While the United Kingdom established the world's first ETS for the six main GHGE in 2002 (see above), Denmark had already adopted a domestic ETS in 1999, although the Danish scheme covered only CO_2 emissions from power stations (Wurzel *et al.* 2013, p. 158). Further evidence for Danish exemplary leadership is the fact that Denmark reduced its GHGE by 31.3% between 1990 and 2015 (EEA 2017). In 2016, 56% of the Danish electricity supply came from renewables. Denmark has the highest share (about 50%) in Europe of combined heat and power production (CHP). Cooperatives and local communities have played a strong role in Denmark's clean energy transition. Together with Germany, Denmark also has the highest share of wind power investment from local cooperatives (Bouwens *et al.* 2016) and was the first country to create a lead market for wind power, thus showing structural leadership.

Already in 2003, Denmark was an early, successful exporter of clean energy technology, which amounted to approximately €4 billion (Hvelplund 2005).

EU multilevel climate governance

The adoption of national climate policy innovations has often constituted the first step in a Europeanisation process that has involved the diffusion of innovations across member states including the subnational level. Examples include the German renewable energy law (*Erneuerbare Energien Gesetz* (EEG)), the United Kingdom's ETS and Denmark's energy-efficiency labels. Such diffusions of climate policy innovations have often taken the form of 'negotiated transfer' (Bulmer and Padgett 2005), which has usually resulted in modifications of the leader's original innovation when it was adopted and implemented by followers. The EU may have influenced even the original national climate innovation by the leader as we can see, for example, in the German EEG which had to be modified following concerns of the Commission about the incompatibility of the draft German EEG with EU competition law.

Since the early 1990s, the EU Commission has tried to directly facilitate subnational climate innovations within the EU's multilevel climate governance system. Initially, the EU established direct links between the EU Commission and local governance actors in regional policy (Marks 1993). As we argue below, climate policy innovation and its diffusion have become a more general phenomenon because the EU has developed 'systemic' opportunity structures for it. An important EU climate change initiative directed at the city level is the Covenant of Mayors, which the EU Commission launched in 2008 (Domorenok forthcoming). It was extended to the global level in 2015. The Covenant, which receives significant EU funding, contains a benchmark of excellence, which supports both exemplary leadership and lesson-drawing. Such institutional arrangements have arguably helped to create a framework for interactive learning at different levels of the multilevel EU climate governance system (Bulkeley and Betsill 2005, see also Kern 2019, this volume).

Subnational leadership

In the EU's multilevel climate governance system, climate-related policy innovation and investment at the local governance level is becoming increasingly important (Jänicke and Quitzow 2017, see also Kern 2019, Wurzel et al. 2019, both this volume). In this section, we, therefore, assess the role of subnational climate leadership.

Germany

Germany has a federal political system in which the states (*Länder*) have frequently offered climate leadership. Progressive German states often influenced both their local communities and the national level, for example, via the Bundesrat. In 1985, Hesse became the first German state with an Environment Minister from the Green Party; Hesse's Environment Minister, Joschka Fischer, encouraged his ministry to become an influential player for the provision of knowledge on the national energy transition (*Energiewende*) in Germany (Krause et al. 1980). In other words, under Fischer's structural leadership, the Hesse Environmental Ministry tried to provide cognitive and exemplary leadership. Subsequent conservative state governments in Hesse put the brake on the rapid expansion of renewable energy. However, the 2012 Hesse Energy Future Law introduced the goal of supplying 100% power and heat from renewables by 2050. Between 1995 and 2013, Hesse achieved GHGE reduction of 24% (HMWEVL 2015). The renewable energy sector generated more than 20,000 jobs by 2013. While Hesse has tried to resume the former *Energiewende* approach under a 'Black-Green' (CDU-Greens) coalition government, elected in 2014, its capital, Frankfurt am Main, has long had a strong 'green' tradition.

Another important German state with innovative climate policies is Baden-Württemberg, which has offered a conservative variant of ecological modernisation since the 1990s. Cities in Baden-Württemberg such as Freiburg and Heidelberg have acted as exemplary climate leaders, which have had a strong innovative influence in Germany and beyond. After the 2011 Fukushima nuclear catastrophe, Baden-Württemberg elected the first Green Prime Minister (*Ministerpräsidenten*) of a German state. Around that time, the Green Party became a coalition partner in the majority of German state governments, which significantly increased its influence in the federal Upper House (*Bundesrat*). Baden-Württemberg introduced an ambitious Act Governing the Mitigation of Climate Change in 2013. In 2016, the Green Prime Minister was re-elected, although this time a 'Green-Black' (Greens-CDU) coalition government succeeded the 'Green-Red' (Greens-SPD) coalition (Jänicke 2017b).

Freiburg is a climate leader because it was one of the first cities in Germany to adopt an energy transition (*Energiewende*). Freiburg regards itself as a prominent example for the climate-friendly transformation of a city (Haag and Köhler 2012). Already in 1986 – the year of the Chernobyl nuclear accident – Freiburg adopted an Energy Supply Concept that demanded the phasing-out of nuclear energy and a significant reduction of CO_2 emissions. In 1996, Freiburg set itself a CO_2 reduction target of 25% by 2010. There is also a strong focus on energy efficiency in all sectors of the city with a steady reduction of final energy consumption. In 2000, work started on Freiburg's car-free Vauban settlement, which features 59 so-called plus-energy buildings and one plus-energy office building. The city has built Vauban as a model district for sustainable living (Müller 2014).

In Bavaria, *Munich* plans to reduce its CO_2 emissions by 50% of 1990 levels by 2030 (Heinelt and Lamping 2015). Munich has a broad spectrum of ambitious mitigation and adaptation activities including a programme for energy-efficient building envelopes and heating renovation, an energy-efficiency of trade initiative, an Eco-Profit programme and a climate-related city map (Covenant of Mayors 2017). Munich is a relatively large and prosperous city with a wide hinterland that has offered not only exemplary but also structural leadership for the surrounding region.

Climate leader villages also play an important role in Germany. Examples include the pioneer villages of Wildpoldsried and Großbardorf in Bavaria and the 'bio-energy village' Jühnde in Lower Saxony. Such villages are leaders because they try to attract followers (see Liefferink and Wurzel 2017) by influencing a broad movement, namely the so-called 100% Renewable-Energy Regions in Germany which collectively represented about 25 million inhabitants (i.e. more than one-quarter of the total population in Germany) in 2014. These villages adopt bottom-up exemplary leadership while experimenting with novel, innovative solutions and expert training (cognitive leadership) and networking (entrepreneurial leadership) which both the German government and EU have supported financially. At first sight, this seems in line with polycentric governance concepts (Ostrom 2010, 2014), which consider bottom-up self-governing initiatives to be more effective than top-down government approaches. However, many of these local climate innovations would not have succeeded without significant funding from 'higher governance' levels (the German federal government and/or the EU).

Because the highly ambitious plans of such villages may lead to their complete reconstruction, we can identify a transformational leadership style. For example, the village of Wildpoldsried, which received the European Energy Award in 2010 and 2014, has a broad spectrum of innovative activities including ambitious renewables and energy saving goals as well as the leasing of e-mobiles. Wildpoldsried's electric power supply from wind, biogas and PV amounted to 688% of the village's own electricity demand in 2016. Wildpoldsried, which started its ambitious climate and environmental policy process in 1997, created at least 140 jobs due to such climate policy–related activities (Wildpoldsried 2017).

The bio-energy village Jühnde in Lower Saxony is a leader in decentralised clean energy supply based on a cooperative model. It provides electricity and heat from bio-energy. Clean power supply exceeds local demand by about 200%. E-mobility is part of the project. Jühnde is also a member of a number of international networks that actively support the visibility of its case. According to Niemann (2015), this 'bio-energy village' has at least 120 followers who have tried to draw lessons from Jühnde.

The United Kingdom

Climate leader regions include Scotland, which has a 100% renewable power target by 2020 and thus acts as an exemplary climate leader within the United Kingdom and beyond. The goal of full decarbonisation of the power sector has been set for 2032. The installed capacity of renewable electricity increased from 2673 MW to 7756 GW between 2007 and 2015, while the share of renewables in power generation amounted to 59% in 2015. This sector created 21,000 jobs (Scottish Renewables 2017).

Climate leader cities include London, which has ambitions to become 'a world leader in tackling climate change' (Greater London Authority 2016). In 2007, the Mayor launched the first Climate Change Plan for London (Greater London Authority 2007). London has set a CO_2 emission reduction target of 60% by 2025 (compared to 1990). The target requires an investment of 40 billion pounds, for example, for urban greening and climate roofing of buildings. London (together with Bogota) created a network of 26 cities which have all signed the C40 Clean Bus Declaration that aims at a 25% share of clean busses by 2020. As London is by far the United Kingdom's largest city, its exemplary and structural leadership potential is considerable, for other UK cities as well as its immediate surrounding regions.

Manchester, which has a CO_2 reduction target that surpasses the national target (Covenant of Mayors for Climate and Energy 2017), has focused strongly on climate protection in its industrial policy. Its low-carbon economic growth sector amounts to a market value of about 4.2 billion pounds and employs over 34,000 people. In 2009, the city region became the first Low-Carbon Economic Area for the built environment, thus showing exemplary and structural leadership (Thorpe 2012).

Climate leader villages in the United Kingdom include Ashton Hayes (1000 inhabitants), which reduced its CO_2 emissions by 24% within 10 years and aims to become carbon neutral. Measures that these villages have taken include the installation of renewable power (mainly PV), improved energy efficiency of buildings and clean energy heating. Schlossberg (2016) has reported lesson-drawing from climate activities in Ashton Hayes by local communities in other countries. The Cornish village of Delabole installed the first commercial wind farm inspired by Danish examples in 1991. Since 2002, Delabole's wind farm has paid about 10,000 pounds sterling annually to the village (Guardian 2017).

Denmark

Copenhagen's 2025 Climate Plan has the objective of turning the city into the world's first carbon neutral capital by 2025. The plan also aims to generate 'employment and green growth' (City of Copenhagen 2012).

Between 2005 and 2011, the city had already reduced its CO_2 emissions by 21%. Aarhus also aims to become carbon neutral, although not until 2030 and since 2008 has adopted several climate action plans (Aarhus Kommune 2017). Thus, comparatively, Copenhagen has shown a higher degree of exemplary leadership.

Denmark has linked its climate and energy strategy strongly to a process of decentralisation for both energy generation and ownership. From the inception of this strategy, local-level actors have played an important role. Already by 1992, these actors installed more than 3000 wind turbines owned by cooperatives (Reiche 2005). Citizen cooperatives have remained important players (Jänicke and Quitzow 2017). The small island of Samsoe (4000 inhabitants), a well-known clean energy leader with strong international connections, within 10 years achieved an energy surplus based on renewable energy (Lewis 2017).

Explaining the 'leaderless leader' paradox in EU climate governance

Most observers have argued that the EU is a global climate governance leader (e.g. Schreurs and Tiberghien 2007, Oberthür and Kelly 2008, Jordan et al. 2010, Wurzel et al. 2017) whose 'climate policy activities have enormous relevance well beyond European borders' (Rayner and Jordan 2013, p. 1). The EU is most of all an exemplary leader that surpasses other regions regarding GHGE reductions and has established a high level of new renewable power capacity. Between 1990 and 2016, the EU achieved a 22.6% reduction of GHGE (EEA 2017). Renewable energy accounted for 86% of the new power capacity added in the EU in 2016, compared with 57% in 2008 (REN21 2017, p. 34). We cannot sufficiently explain this type of exemplary leadership merely by top-level EU climate policy measures and decisions. The EU ETS, which is the EU's core climate policy instrument (Eikeland and Skjaerseth 2019), has remained largely ineffective (Jänicke and Quitzow 2017). In 2017, the EU ETS's carbon price was about €5/ton of carbon and thus had little effect on corporate actors' decisions. The EU was also not relying on strong, harmonised instruments to stimulate green electricity (Jacobs 2012). Moreover, financial support for renewable energy has diminished significantly in most member states and at the EU level in recent years.

Scholars have identified a 'leaderless leader' paradox whereby 'the EU seeks to lead by example but is itself a relatively leaderless system of governance' (Jordan et al. 2012, p. 6). This helps to explain the discrepancy between top-down EU climate policy measures, which are relatively modest, and the actual achievements as regards the reduction of GHGE and the increase in renewable power capacity. Schreurs and Tiberghien (2007, p. 22) have tried to resolve the leaderless leader paradox by arguing that 'EU leadership in climate policy is the result of the dynamic process of competitive multi-level reinforcement among different political poles within a context of decentralised governance'. Multilevel reinforcement, especially between the member state and EU levels of governance constitutes an important explanatory factor for the EU's relatively ambitious climate policies. However, in recent years, the EU's multilevel system has also developed a strong subnational governance dimension, which has remained under-researched.

Within the complex EU climate governance system, a relatively wide range of actors are involved in 'baton passing' at different governance levels (Schreurs and Tiberghien 2007, p. 24). Scholars have paid significant attention to how the climate leaders among the EU's member states (such as Denmark, Germany and the United Kingdom) and EU institutional actors (e.g. the European Parliament (EP) and Commission) have tried to influence climate governance at the EU and member states levels (Schreurs and Tiberghien 2007, Oberthür and Kelly 2008, Jordan et al. 2010, Wurzel et al. 2017, Matschoss and Repo 2018). Although there is a growing literature on the role of cities and city networks (Kern and Bulkeley 2009, Betsill and Bulkeley 2013, Eckersley 2018), scholars know relatively little about how subnational (and societal) actors are both affected by and affect the EU's multilevel climate governance system.

There are at least four main reasons for the emerging interest in subnational climate governance innovations within the EU multilevel climate governance system. First, cities are both major greenhouse gas (GHG) emitters *and* laboratories for innovative climate governance measures, some of which could be scaled-up to 'higher' levels of governance (Betsill and Bulkeley 2013, see also Wurzel *et al.* 2019, this volume). Second, especially since the 2008 financial crisis, the EU Commission has pushed its 'better regulation' agenda of adopting top-down direct regulation only when necessary, which is broadly in line with the principle of subsidiarity already adopted in the 1993 Maastricht Treaty. Third, the international multilevel climate governance system has become more polycentric with the 2015 Paris Agreement (Wurzel et al. 2017, Oberthür 2018). Fourth, as already discussed above, the EU multilevel climate governance system has also increasingly exhibited polycentric features such as EU support for city networks and the Covenant of Mayors.

We could characterise the EU's polycentric climate governance features (Ostrom 2010, 2014) as a 'multi-impulse mechanism' (Klemmer et al. 1999, Jänicke 2017a).[3] The multi-impulse mechanism concept focuses on the governance *effects* (i.e. not the structure) of the polycentric features of MLG systems such as the EU. 'Impulse' in this context essentially means an external stimulus or *impetus to learn*. Lesson-drawing from leaders in this multi-impulse system often takes place within transnational networks. It can be the result of cooperation as well as competition. It is achievable by vertical up-scaling from best practice at lower levels or top-down climate policy decisions and policies. In other words, it should be possible to observe all of the above-mentioned four main driving factors – lesson-drawing, convergence, transnational networks and competition – for climate innovation within the EU multilevel climate governance system. Here, we have focused primarily on lesson-drawing which, under certain circumstances, can develop into *a dynamic system of interactive learning* within the EU multilevel climate governance system.

In contrast to many polycentric climate governance approaches (e.g. Ostrom 2014), we argue that top-level policy decisions and the EU's institutional 'infrastructure' strongly influence climate leadership dynamics and lesson-drawing within the EU multilevel climate governance system, encouraging climate innovation at different governance levels. A system of multilevel interactive learning has emerged which is neither leaderless nor merely the result of bottom-up processes. Of central importance for the learning process is the recognition that economic co-benefits (e.g. employment, innovation and productivity) can result from climate governance measures (Mayrhofer and Gupta 2016). All levels of the EU's multilevel climate governance system have learned lessons about economic co-benefits of climate governance measures (Jänicke and Quitzow 2017). It is due to the economic co-benefits of EU climate governance measures that veto players (Tsebelis 2002) and 'joint decision traps' (Scharpf 1988) have not prevented the EU from acting as a climate leader (Oberthür and Kelly 2008, Wurzel et al. 2017). EU climate policy as a 'business case' has become a success story overall. However, especially the Visegrad countries (Hungary, Poland, Czech Republic and Slovakia) have remained sceptical about the move towards a low-carbon economy (Skjaerseth 2018) and the concept of ecological modernisation, which assumes that ambitious environmental measures are beneficial for both the environment and economy.

While the integration of general environmental requirements into other policy areas – often referred to as environmental policy integration (EPI) – has made little progress on the EU level (Jordan and Lenschow 2008), the EU seems to have achieved a better record with regard to the integration of climate policy concerns into non-climate policy areas such as regional policy and budgetary policy. This is not to argue that climate policy integration (CPI) has been successfully achieved for all EU policies. There are EU policies (e.g. transport and agriculture) for which little, if any, meaningful CPI has occurred up to now (Jordan et al. 2012, p. 58, Dupont and Oberthür 2015).

EU support for multilevel climate governance

The EU, and in particular the European Commission, have actively advanced initiatives (e.g. the Covenant of Mayors) and mechanisms (e.g. regional policy funding) to mobilise local governance actors with the aim of enabling them to develop and showcase their climate governance innovations. Non-EU states such as the United States, China and India also have a MLG approach to climate protection (Wurzel et al. 2017). However, the architecture of the EU's multilevel climate governance system is comparatively more advanced, particularly because EU institutional arrangements and financial mechanisms support lower climate governance levels (Jänicke and Quitzow 2017).

The EU's first institutionalised MLG innovations, which directly targeted the subnational level, occurred in regional policy (Marks 1993). These MLG regional policy initiatives, which predate EU climate policy, required time before effective institutional arrangements for lesson-drawing from subnational leaders could be set up (Marks and Hooghe 2004). The EU's regional policy contains a strong financial commitment for the shift towards a low-carbon economy. A total of 172 regions accounting for 80% of the EU regions participated in the Smart Specialisation Platform on Energy. The platform helps regions to share their expertise on sustainable energy investments and especially on the deployment of innovative low-carbon technologies (CEU 2015). Although the Committee of the Regions (CoR) has few formal powers, it became an important institution for the exchange of ideas and practical experience gained with climate innovation at the regional and local governance levels. In 2014, the Committee published a Charter for Multilevel Governance (MLG) in Europe (CoR 2014).

Even more important was the support for cities from funding mechanisms such as the European Structural and Investment Funds (ESIF) and the incentives offered by the European Green Capital Awards. While such mechanisms indirectly support climate mitigation activities, the Covenant of Mayors has been explicitly linked to the EU's 2008 climate and energy package (Bendlin et al. 2016). The Commission launched the Covenant of Mayors, which was integrated into the Covenant of Mayors for Climate and Energy in 2015, to facilitate local climate innovations. With its benchmark of excellence, the Covenant of Mayors provided a significant institutional stimulus for local exemplary climate leadership and lesson-drawing. By 2017, it had attracted 7675 signatories from local communities (including some non-EU cities), representing 241 million inhabitants. In total, 5992 of these local communities and cities have Action Plans with 2020 targets, although there is a new objective of achieving at least 40% GHGE reductions by 2030 (Covenant of Mayors for Climate and Energy 2017). The average targets of these Actions Plans, which amount to −27% of CO_2

emissions, surpass the EU's collective GHGE reduction targets for 2020 (Covenant of Mayors, 2017). Importantly, the EU has linked the Covenant of Mayors for Climate and Energy (together with the Compact of Mayors) to the global level with the adoption of the Global Covenant of Mayors for Climate and Energy (2018) in which9149 cities representing 781 million people worldwide participate. The EU's exemplary leadership has, therefore, attracted followers at the local governance level also in other parts of the world.

The EU has extended its multilevel climate governance system through the Covenant of Mayors by broadening it to the village level which has become important because renewable energy investment often takes place in rural areas, most of which also provide the necessary sinks (e.g. forests) for 'negative emissions'. The EU Commission has started a pilot project on smart 'eco-social villages', which uses best practice for a more general approach to rural development (CEU 2017).

Another actor that has strengthened the EU multilevel climate governance system has been the European Investment Bank (EIB), which has used financial instruments to offer 25% of its credits for climate related investment (EIB 2017). Since the early 2010s, the EU must spend 20% of its budget on climate-related measures. In 2016, Commission President Juncker announced in his 'State of the Union' address that, under the European Fund for Strategic Investment (EFSI), 40% of large infrastructure and innovation projects have to contribute to climate action, although observers later criticised some of the investment for including fossil-fuel projects (ENDS Europe 2017).

The EU's attempt to move towards an *Energy Union* is another institutional mechanism, although progress has been slow and patchy. The EU initially introduced Energy Union without explicit climate policy considerations but eventually rectified this, for instance, by making 'efficiency first' an important goal (Verhaar and Frassoni 2017). Moreover, the importance of subnational and societal actors has been recognised in the European Commission's 2016 Communication Accelerating Clean Energy Innovation, which stated that 'the transition to a low-carbon, energy-efficient and climate-resilient economy, will require a more decentralised, open system with involvement of society' (CEU 2016, p. 4).

In short, the EU has initiated and/or supported a large number of subnational climate governance initiatives, which have made the EU climate governance system arguably more polycentric. However, many decisions on funding and GHGE reduction targets, which have an indirect effect on subnational climate governance innovations, are taken at the top level of the EU climate governance system. Therefore, the EU can offer exemplary climate leadership even without direct interference at the local level.

Based on the evaluation of 262 EU climate policies, Berkhout and colleagues (2010, p. 137) have concluded that 'climate policy is deeply multi-level, but with a trend towards harmonisation at the EU level'. Jörgens and Solorio (2017) have referred to it as bottom-up Europeanisation, which they distinguish from both top-down Europeanisation and horizontal Europeanisation (see also Jordan and Liefferink 2004). Horizontal Europeanisation refers not only 'to the direct diffusion or transfer of policies from one EU member state to another' (Jörgens and Solorio 2017, p. 11). Instead, horizontal Europeanisation in the form of broad lesson-drawing from best practice takes place at all levels of the EU's multilevel climate governance system, including the provincial level and the local level (Kern and Bulkeley 2009, Jänicke and Quitzow 2017, see also Kern 2019, Wurzel *et al.* 2019, this volume).

Conclusion

National leaders have long played an important role in developing EU environmental policy in general (e.g. Andersen and Liefferink 1997) and climate policy in particular (e.g. Oberthür and Kelly 2008, Jordan *et al.* 2012, Wurzel *et al.* 2017). The opening up of windows of opportunity unrelated to climate governance has sometimes enabled their leadership. For example, Germany's climate policy benefitted significantly from 'wall fall profits' and the United Kingdom from its 'dash for gas'. Especially since the early 1990s, the EU system of multilevel climate governance incrementally developed *systemic* opportunity structures that have encouraged climate leadership and lesson-drawing at different governance levels, including the regional and city levels. Economic co-benefits resulting from climate mitigation measures have provided attractive economic exemplary leadership examples from which other actors have drawn lessons.

The literature has identified a 'leaderless leader' paradox according to which the EU has become a leader in global climate governance (e.g. Oberthür and Kelly 2008, Wurzel *et al.* 2017) although it is itself a relatively leaderless system. We cannot sufficiently explain the EU's global exemplary leadership merely by its top-level climate policy. We have argued that the following two main factors can best explain the 'leaderless leader' paradox: the use and purposeful extension of the EU multilevel climate governance system into a system which encourages interactive climate policy learning at all climate governance levels including the subnational level and the recognition of the economic co-benefits of climate governance measures and the integration of climate policy objectives into non-climate policies such as budgetary policy.

More research is necessary to improve our understanding of the exact impact that the EU multilevel climate governance system has on subnational climate governance. The role of peer-to-peer learning at different governance levels (in particular at the regional and/or local governance levels) remains under-researched. The interactions between different climate governance levels also require urgently additional scholarly attention. Moreover, there is a lack of research into climate policy failures, especially at the subnational climate governance level of the EU multilevel climate governance system. While there is a growing literature on relatively affluent climate leader cities (e.g. Kern and Bulkeley 2009, Kern 2019, this volume) relatively little is known about climate innovations in deprived, structurally disadvantaged cities (see however Jonas et al. 2017, Wurzel et al. 2019, this volume) and how such cities can affect the EU's multilevel climate governance system and vice versa.

Notes

1. *Leadership by example* has a long tradition in European history. For example, French absolutism's power structure, economic system, architecture and even its preferred products have been imitated by other European countries.
2. Liefferink and Wurzel (2017) and Wurzel et al. (2017) have argued that leaders actively seek to attract followers while this is not normally the case for pioneers. Here, we focus primarily on leaders.
3. Klemmer et al. (1999) first used the term 'multi-impulse-hypothesis' for environmental innovations that are not caused by one specific policy instrument but by the interactions of different societal factors.

Acknowledgments

Martin Jänicke delivered an early version to the Innovation in Global Climate Governance (INOGOV) funded workshop on 'Pioneers and Leaders in Polycentric Climate Governance (PiLePoC)' in Hull, 15–16 September 2016. Both authors are grateful to the referees and the journal editors, Chris Rootes and Anthony Zito, for their very helpful comments.

Disclosure statement

No potential conflict of interest was reported by the authors.

ORCID

Rüdiger K.W. Wurzel http://orcid.org/0000-0001-5873-4232

References

Andersen, M.S. and Liefferink, D., eds., 1997. *European environmental policy. the pioneers*. Manchester and New York: Manchester University Press.

Bendlin, L., et al. 2016. Cities views and ownership of the Covenant of Mayors. In: J. Kemmerziell, ed. *Städte und Energiepolitik im europäischen Mehrebenensystem*. Baden-Baden: Nomos, 103–124.

Berkhout, F., et al. 2010. How do climate policies work? In: M. Hulme and H. Neifeldt, eds. *Making climate change work for us*. Cambridge: Cambridge University Press, 137–164.

Betsill, M. and Bulkeley, H., 2013. Revisiting the urban politics of climate change. *Environmental Politics*, 22 (1), 136–154. doi:10.1080/09644016.2013.755797

Bouwens, T., Gotchev, B., and Holstenkamp, L., 2016. What Drives the Development of Community Energy in Europe? *Energy Research and Social Science*, 13, 136–147. doi:10.1016/j.erss.2015.12.016

Bulkeley, H., and Betsill, M., 2005. Rethinking sustainable cities: multilevel governance and the 'urban' politics of climate change. *Environmental Politics*, 14 (1), 42–63.

Bulmer, S., and Padgett, S., 2005. Policy transfer in the European Union: an institutionalist perspective. *British Journal of Political Science*, 35 (1), 103–126.

CEU, 2015. *Smart specialisation platform on energy* [online]. Brussels: Commission of the European Union. Available from: https://ec.europa.eu/jrc/en/news/new-smart-specialisation-platform-on-energy-launched [Accessed 31 Jan 2018].

CEU, 2016. *Communication from the commission. Accelerating climate energy innovation. COM(2016) 763 final*. Brussels: Commission of the European Union.

CEU 2017. Smart eco-social villages for rural development. Brussels: commission of the European Union. Available from: https://ec.europa.eu/info/news/smart-eco-social-villages-rural-development-2017-apr-04_en [Accessed 31 January 2018].

City of Copenhagen, 2012. *CPH 2025 climate plan*. Copenhagen: City of Copenhagen.

CoR (Committee of the Regions), 2014. *On the charter for multilevel governance in Europe*. (RESOL-V-=12). Available from: http://cor.europa.eu/en/activities/governance/Documents/mlg-charter/en.pdf [Accessed 31 January 2018].

Covenant of Mayors for Climate and Energy, 2017. Key Figures. Available from: https://ec.europa.eu/jrc/en/publication/covenant-mayors-climate-and-energy-default-emission-factors-local-emission-inventories-version-2017 [Accessed 31 January 2018]

Deutscher Bundestag, 1990. *Schutz der Erde*. Bonn: Deutscher Bundestag.

Domorenok, E., forthcoming. Voluntary instruments for ambitious objectives? The experience of the EU Covenant of Mayors. *Environmental Politics*.

Dupont, C. and Oberthür, S., eds., 2015. *Decarbonization in the European Union*. London: Routledge.

Eckersley, P., 2018. Who shapes local climate policy? Unpicking governance arrangements in English and German cities. *Environmental Politics*, 27 (1), 139–160. doi:10.1080/09644016.2017.1380963

EEA, 2017. *Annual European Union greenhouse gas inventory 1990–2015 and inventory report 2017*. Copenhagen: European Environment Agency.

EIB (European Investment Bank) 2017. *Klima und Umweltschutz für unsere Zukunft*. Available from: http://www.eib.org/projects/priorities/climate-and-environment/index.htm [Accessed 31 January 2018].

ENDS Europe, 2017. Revamped juncker plan to target climate action. *ENDS Europe*, 14 September.
Global Covenant of Mayors for Climate and Energy 2018. Available from: http://www.globalcovenantofmayors.org/ [Accessed 31 January 2018]
Greater London Authority, 2007. *Today to protect tomorrow: the Mayors climate action plan*. CCAP-2007(1). London: London Greater Authority.
Greater London Authority, 2016. *The London Plan 2016*. London: Greater London Authority. Available from: https://www.london.gov.uk/sites/default/files/the_london_plan_2016_jan_2017_fix.pdf [Accessed 3 April 2018].
Guardian, 2017. Cornish village marks 25 years of UK wind power, *The Guardian*. Available from: https://www.theguardian.com/.../cornish-village-delabole-25-years-of-uk-wind-power [accessed 4 April 2018].
Haag, M. and Köhler, B., 2012. Freiburg im Breisgau – nachhaltige Stadtentwicklung mit Tradition und Zukunft. *Informationen zur Raumentwicklung*, 5/6, 243–256.
Hayward, J., ed., 2008. *Leaderless Europe*. Oxford: Oxford University Press.
Heinelt, H. and Lamping, W., 2015. *Wissen und entscheiden: lokale Strategien gegen den Klimawandel in Frankfurt am Main*. München: Campus-Verlag.
Héritier, A., Knill, C., and Mingers, S., 1996. *Ringing the changes in Europe*. Berlin: Walter de Gruyter.
HMWEVL, 2015. *Energiewende in Hessen. Monitoringberichtt 2015*. Wiesbaden: Hessisches Ministerium für Wirtschaft, Energie, Verkehr und Landesentwicklung.
Hvelplund, F., 2005. Denmark. *In*: D. Reiche, ed. *Handbook of renewable energies in the European Union*. Frankfurt/M.: Peter Lang, 83–100.
Jacobs, D., 2012. *Renewable energy convergence in the EU*. Surrey: Ashgate.
Jänicke, M. and Weidner, H., 1997. Germany. *In*: M. Jänicke and H. Weidner, eds. *National environmental policies*. Berlin: Springer, 133–155.
Jänicke, M., 2005. Trend setters in environmental policy: the character and role of pioneer countries. *European Environment*, 15 (2), 129–142. doi:10.1002/eet.v15:2
Jänicke, M., 2017a. The multi-level system of global climate governance – the model and its current state. *Environmental Policy and Governance*, 27, 108–121. doi:10.1002/eet.1747
Jänicke, M., 2017b. Germany: innovation and climate leadership. *In*: R. Wurzel, J. Connelly, and D. Liefferink, eds. *The European Union in international climate change politics. Still taking a lead?* London: Routledge, 114–129.
Jänicke, M. and Quitzow, R., 2017. Multi-level reinforcement in European climate and energy governance: mobilizing economic interests at the sub-national levels. *Environmental Policy and Governance*, 27, 122–136. doi:10.1002/eet.1748
Jonas, A., *et al.*, 2017. Climate change, the green economy and re-imagining the city. *Die Erde*, 148 (4), 197–211.
Jordan, A., *et al.*, 2003. European governance and the transfer of New Environmental Policy Instruments (NEPIs) in the European Union. *Public Administration*, 81 (3), 555–574. doi:10.1111/1467-9299.00361
Jordan, A., *et al.*, 2012. Understanding the paradoxes of multilevel governing: climate change policy in the European Union. *Global Environmental Politics*, 12 (2), 43–66. doi:10.1162/GLEP_a_00108
Jordan, A. and Lenschow, A., eds., 2008. *Innovation in environmental policy?* Cheltenham: Edward Elgar.
Jordan, A. and Liefferink, D., eds., 2004. *Environmental policy in Europe: the Europeanisation of national policy*. London: Routledge.

Jordan, A., et al., eds., 2010. *Climate change policy* in the European Union. Cambridge: Cambridge University Press.

Jordan, A., Wurzel, R., and Zito, A., 2013. Still the century of 'new' environmental policy instruments? Taking stock and exploring the future. *Environmental Politics*, 22 (1), 155–173. doi:10.1080/09644016.2013.755839

Jörgens, H. and Solorio, I., 2017. The EU and the promotion of renewable energy. *In*: I. Solorio and H. Jörgens, eds. *A guide to European renewable energy policy*. Cheltenham: Edward Elgar, 3–22.

Kern, K., 2019. Cities as leaders in EU multi-level climate governance? Embedded upscaling of local experiments in Europe. *Environmental Politics*, 28 (1).

Kern, K. and Bulkeley, H., 2009. Cities, Europeanization and multi-level governance. *Journal of Common Market Studies*, 47 (2), 309–332. doi:10.1111/j.1468-5965.2009.00806.x

Klemmer, P., Lehr, U., and Löbbe, K., 1999. *Umweltinnovationen – anreize und Hemmnisse*. Berlin: Analytica.

Krause, F., et al., 1980. *Energie-Wende. Wachstum und Wohlstand ohne Erdöl und Uran*. Frankfurt: S. Fischer.

Lewis, D., 2017. Energy positive: how Denmark's Samsoe island switched to zero carbon. *The Guardian*, 23 Feb.

Liefferink, D. and Wurzel, R.K.W., 2017. Environmental leaders and pioneers: agents of Change? *Journal of European Public Policy*, 24 (7), 951–968. doi:10.1080/13501763.2016.1161657

Marks, G., 1993. Structural policy and multi-level governance in the EC. *In*: A. Cafruny and G. Rosenthal, eds. *The state of the European community. Volume 2*. Boulder, Colorado: Lynne Reiner, 391–411.

Marks, G. and Hooghe, L., 2004. Contrasting visions of multilevel governance. *In*: I. Bache and M. Flinders, eds. *Multilevel governance*. Oxford: Oxford University Press, 15–30.

Matschoss, K. and Repo, P., 2018. Governance experiments in climate action: empirical findings in the 28 European Union countries. *Environmental Politics*, 27 (4), 598–620. doi:10.1080/09644016.2018.1443743

Mayrhofer, J.P. and Gupta, J., 2016. The science and politics of co-benefits in climate policy. *Environmental Science and Policy*, 57, 22–30. doi:10.1016/j.envsci.2015.11.005

Meyer, N.I. and Koefoed, A.L., 2003. Danish energy reform. *Energy Policy*, 31, 597–607. doi:10.1016/S0301-4215(02)00145-3

Müller, H., 2014. Plusenergiehaus: freiburgs Solarsiedlung als Vorzeigeprojekt, *Die Welt*, 30 March.

Niemann, H., 2015. Das Fortschrittsdorf Jühnde. *Göttinger Tageblatt*, 20 Sep.

Oberthür, S., 2018. Reflections on global climate politics post Paris. Power, interests and polycentricity. *The International Spectator*, 51 (4), 80–94. doi:10.1080/03932729.2016.1242256

Oberthür, S. and Kelly, C.P., 2008. EU leadership in international climate policy. *The International Spectator*, 43 (3), 35–50. doi:10.1080/03932720802280594

Ostrom, E., 2010. Polycentric systems for coping with collective action and global environmental change. *Global Environmental Change*, 20, 550–557. doi:10.1016/j.gloenvcha.2010.07.004

Ostrom, E., 2014. A polycentric approach for coping with climate change. *Annals of Economics and Finance*, 15 (1), 97–134.

Radaelli, C., 2000. Policy transfer in the EU. *Governance*, 13 (1), 25–43. doi:10.1111/0952-1895.00122

Rayner, T. and Jordan, A., 2017. The United Kingdom; A record of leadership under threat. *In*: R. Wurzel, J. Connelly, and D. Liefferink, eds. *The European Union in international climate change politics*. London: Routledge, 173–188.

Rayner, T. and Jordan, A., 2013. The European Union: the polycentric climate policy leader? *Wiley Interdisciplinary Reviews: Climate Change*, 4, 75–90. doi:10.1002/wcc.205

Rehbinder, E. and Stewart, R., 1985. *Integration through law. Europe and the American federal experience*. Berlin: Walter de Gruyter.

Reiche, D., ed., 2005. *Handbook of renewable energies in the European Union*. Frankfurt./M.: Peter Lang.

REN21, 2017. *Renewables 2017 – global status report*. Paris: REN21.

Rose, R., 1993. *Lesson-drawing in public policy. A guide to learning across time and space*. Chatham, N.J.: Chatham House.

Scharpf, F.J.W., 1988. The joint-decision trap: lessons from German federalism and European integration. *Public Administration*, 66 (2), 239–278. doi:10.1111/j.1467-9299.1988.tb00694.x

Schlossberg, T., 2016. English village becomes climate leader. *New York Times*, 21 Aug.

Schreurs, M. and Tiberghien, Y., 2007. Multi-level reinforcement: explaining European Union leadership in climate change mitigation. *Global Environmental Politics*, 7 (4), 19–46. doi:10.1162/glep.2007.7.4.19

Scottish Renewables 2017. Renewables in Numbers. Available from: https://www.scottishrenewables.com/sectors/renewables-in-numbers/ [Accessed 15 January 2018]

Skjaerseth, J.B., 2018. Implementing EU climate and energy policies in Poland: policy feedback and reform. *Environmental Politics*, 27 (3), 498–518. doi:10.1080/09644016.2018.1429046

Thorpe, K., 2012. Overview of UK national & local policies & frameworks for low carbon cities and communities. *Presentation at the UK-Taiwan cities forum*, 1 Nov.

Tsebelis, G., 2002. *Veto players. How political institutions work*. Princeton: Princeton University Press.

Verhaar, H. and Frassoni, M. 2017. How can energy union governance help put efficiency first? *EURACTIV*, 9 Jan.

Wildpoldsried, 2017. *Wildpoldsried* [online]. Das Energiedorf. Available from: www.wildpoldsried.de. [Accessed 3 Apr 2018]

Wurzel, R., Liefferink, D., and Torney, D., 2019. Pioneers, leaders and followers in multilevel and polycentric climate governance. *Environmental Politics*, 28 (1).

Wurzel, R.K.W., 2002. *Environmental policy-making in Britain, Germany and the European Union*. Manchester: Manchester University Press.

Wurzel, R.K.W., 2010. Environmental, climate and energy policies: path-dependent incrementalism or quantum leap? *German Politics*, 19 (3–4), 460–478. doi:10.1080/09644008.2010.515838

Wurzel, R.K.W., Connelly, J., and Liefferink, D., 2017. *The European Union in international climate change politics. Still taking a lead?* London: Routledge.

Wurzel, R.K.W., Zito, A., and Jordan, A., 2013. *Environmental governance in Europe*. Cheltenham: Edward Elgar.

Environmental, climate and social leadership of small enterprises: Fairphone's step-by-step approach

Katja Biedenkopf ◉, Sarah Van Eynde and Kris Bachus ◉

ABSTRACT
Achieving sustainable consumption and production requires a break with current practices in many sectors, including the smartphone sector. Leaders are central actors in catalysing such change by developing, implementing and promoting innovative ideas, products and practices. Not only large but also small enterprises can aspire to assume leadership for sustainability. This contribution explores the environmental, climate and social leadership of the social enterprise *Fairphone* that seeks to start a movement towards a more sustainable smartphone sector. Endowed with barely any structural power, it relies on other leadership types, especially entrepreneurial leadership, which is based on dialogue, persuasion and coalition-building. Small enterprises can be leaders, but pursuing a goal such as transforming the smartphone sector takes a step-by-step approach targeting different follower groups from suppliers, competitors and consumers to end-of-life processors and policymakers. Those different follower groups are susceptible to different (combinations of) leadership types.

Introduction

Achieving sustainable consumption and production requires a break with current practices in many sectors. Leaders are central actors in catalysing such change by developing, implementing and promoting innovative ideas, products and practices. Small enterprises can be leaders, but pursuing a goal such as transforming the smartphone sector takes a step-by-step approach targeting different follower groups from suppliers, competitors and consumers to end-of-life processors and policymakers.

Multinational companies have become critical players in environmental governance, alongside states and other public authorities (Dupuis and Schweizer 2019 – this volume). Through their globalised production and supply chains, they exert influence on environmental, climate and social conditions (Newell 2001). Although multinationals seem well placed to exert pressure on supply chain actors, their use of codes of conduct and auditing schemes has not yielded sufficient improvements of environmental, climate and social conditions so far (Lund-Thomsen and Lindgreen 2014, Egels-Zandén and Lindholm 2015). Codes of conduct and audits mainly target the direct suppliers of a brand company rather than addressing the supply chain more holistically (Locke and Samel 2012, Nadvi and Raj-Reichert 2015). Moreover, brand companies' leverage has diminished since they have mostly outsourced production to contract manufacturers that produce devices on their behalf. Those contract manufacturers have become large players. They enhanced their capabilities and diversified their customer basis, which sharply reduces large brands' structural leverage (Gereffi et al. 2005, Raj-Reichert 2015). In addition, multinationals' capacity to act as drivers of environmental, climate and social justice has been questioned since business considerations generally seem to motivate them, using sustainability as a business tool rather than pursuing fundamental societal change (Dauvergne and Lister 2013).

Small social enterprises differ in this regard and can play a key role in markets that large brand companies dominate, including the smartphone sector. Social enterprises distinguish themselves from conventional enterprises by their specific hybrid character between non-governmental organisation (NGO) and business actor. Two defining features characterise them: the primacy of social – including environmental and climate – aims and their primary activity of trading goods and services (Peattie and Morley 2008, Dichon and Anderson 2009). The subgroup of social entrepreneurs seizes opportunities for market-changing innovations in the pursuit of an environmental, climate or social goal. Such a social entrepreneur actively strives to create positive change through innovative products or services rather than reactively responds to market forces (Alvord et al. 2004, Luke and Chu 2013). Their explicit goal to initiate environmental, climate and social change makes social entrepreneurs a likely case for studying leadership while their negligible structural power raises the question of how they could possibly exert leadership. This is the core puzzle for this contribution.

Liefferink and Wurzel's (2017) leadership perspective provides a useful tool for understanding small enterprises' various means of influence in global value chains. They propose a nuanced differentiation of leadership types that can enrich existing conceptualisations put forward by the global value chain and social entrepreneurship literature. Although Locke et al. (2009) already identify joint problem solving, information exchange and the diffusion of best practices as mechanisms enhancing the working conditions in global supply chains, Liefferink and Wurzel's (2017) framework offers a more holistic approach. According to them, actors can exert leadership in four distinct ways: by developing innovative ideas and solutions, by implementing ambitious measures internally to establish an example, by convincing others to adopt ambitious practices through diplomatic skills and by using structural power to coerce change (Liefferink and Wurzel 2017). In this contribution, we adjust this leadership framework – originally developed for governmental actors in climate negotiations and policymaking – to non-state actors in global production and consumption networks and apply it to the social enterprise Fairphone.

Fairphone 'started [...] as a campaign aimed at creating more awareness around the abuses in the supply chain of electronics' (van Abel 2013). It pursues the goal of designing smartphones that incorporate the ideas of longevity, ownership and fairness (Hebert 2015b) 'to improve the electronics value chain one step at a time' (Fairphone 2016a). The goal goes beyond producing a more environmentally and socially sustainable phone and includes compelling broader change beyond Fairphone's own operations. Fairphone thus pursues a collective good, which according to Skodvin and Andresen (2006) is one of the characteristics of a leader. The nature of their goal differentiates leaders from other types of actors. Leaders pursue a common goal, whereas other entrepreneurs and negotiators pursue self-interested goals (Malnes 1995).

While small and medium enterprises (SMEs) possess no meaningful structural leverage compared to multinationals, they are more agile, which enables them to engage in experimentation, acting as a laboratory for change. They can cater to an audience willing to pay a higher price for sustainable products, enabling them to pursue more ambitious goals than large brands that cater to mass markets where they largely compete on price. Surprisingly, studies on how SMEs strategically pursue environmental, climate and social goals are scarce, with few exceptions such as Egels-Zandén's (2016, 2017) analysis of the garment company Nudie Jeans.

Changing environmental, climate and social conditions in the smartphone supply chain is challenging. A leader must reach a large number of dispersed actors. The global value chain literature (e.g. Gereffi *et al.* 2005) mainly focuses on brand companies that exert influence upwards on their supply chain, while the global production network literature (e.g. Raj-Reichert 2012) includes NGOs and governmental actors when analysing environmental, climate and social change and stagnation. Our contribution includes an even broader scope of actors by adding end-consumers and end-of-life treatment actors. We conceptualise these actor groups as addressees of leadership activities rather than leaders or influencers in their own right. Our analysis centres on one specific leader: the social enterprise Fairphone. By exposing a company's leadership activities not only towards its direct suppliers but also towards all other actors along the value chain, consumers, end-of-life processors and policymakers, we improve the understanding of the complex mechanisms through which leadership unfolds. Adding the perspective of different follower groups, this contribution adds to the nascent but increasingly important literature on supply chain leadership (Gosling *et al.* 2016).

The next section outlines the sustainability challenges and complexity of the smartphone sector. Then, we tailor the four distinct leadership types to social enterprises in the smartphone sector by differentiating five follower groups. We apply this conceptual framework to the Fairphone case, tracing its distinct leadership activities. The concluding section proposes some areas for future research.

The environmental, climate and social impact of smartphones

Smartphones have a significant ecological, climate and social footprint, encompassing all phases of their lifecycle. The impact on low-income countries of phones sold and used in high-income countries is significant. It mainly consists of bad working conditions and damaging environmental impacts on air, soil and water, including greenhouse gas emissions. The persistence of those problems results from a combination of factors, including lax governmental regulation and enforcement, complex and dispersed global value chains, a lack of capacity and brand company management practices that render component manufacturer compliance difficult (Gereffi *et al.* 2005, Locke and Samel 2012).

Producing sustainable smartphones requires not only addressing a brand company's own practices but also the entire value chain, including environmental, climate and social improvements by suppliers without which a brand company cannot produce truly sustainable products (van Lakerveld and van Tulder 2017). Numerous actors participate in mining, producing or assembling (sub)parts of the final smartphone in various geographical locations. Pursuing ambitious environmental, climate and social goals is highly challenging since each step of the process and each actor faces a variety of environmental, climate and social problems. Locke and Samel (2012) identify the problem that electronics companies often only focus on their direct supplier companies, neglecting the longer supply chain and the varied product usage.

Environmental, climate and social challenges accompany a smartphone's entire life cycle. A device's design phase is decisive for the sustainability of its entire life cycle. Design determines a smartphone's longevity, energy use, reparability and recyclability. The mining phase of raw materials is often problematic. Smartphones contain several metals, including indium, gold, silver, lithium, cobalt, copper, nickel, alumina, tin, lead, tantalum and silicon. In many cases, mining occurs in countries with low environmental, climate and social protection due to the absence or non-enforcement of regulation (UNECA 2011, Moran *et al.* 2014). Moreover, in some regions, the so-called conflict minerals pose a problem (Veale 2012): rebel groups known to be involved in local conflicts extract minerals such as tungsten from Rwanda to finance their activities.

The manufacturing phase creates several environmental, climate and social impacts, often in geographical areas other than the sales market. Energy consumption contributes a substantial share of the manufacturing impact, with significant climate impacts when based on burning fossil fuels (Ercan *et al.* 2016). A large portion of smartphone production is in countries where environmental, climate and worker protection is less strict than in Europe. The environmental and climate impacts include greenhouse gas emissions, and water, soil and air pollution at a local level. The social impact includes poor working conditions, long working hours and low wages. Numerous companies jointly form a complex production chain from foundries and materials processors to component manufacturers of for example integrated circuits, batteries, cameras and displays and contract manufacturers that assemble the final product. Moreover, transportation of components and smartphones contributes to smartphones' climate impact.

In the use phase, consumer behaviour plays a central role. Proper battery management can extend a phone's lifetime. Another important aspect is the premature discarding of devices. In Western countries, the average smartphone use is no longer than 2 years (Sarath *et al.* 2015), which is far below its potential lifespan. Modular design and easy reparability can extend the use phase even further. Also, the disposal phase depends to some extent on the consumer. They may bring end-of-life smartphones to responsible recycling facilities, which can reuse and recycle a number of materials. Europe and many US states mandate appropriate recycling, but other countries do not have mandatory electronic waste provisions. Many devices end up in dumps or suboptimal recycling facilities domestically or are – in Europe illegally – shipped to low-income countries, where primitive methods such as open-air incineration and acid baths are used to extract the materials. Those practices emit dioxins, furans and other pollutants, which harms the often-unprotected workers and contaminates air, soil and water (Greenpeace 2008, Biedenkopf 2015).

The complexity of the policy problems associated with the smartphone life cycle poses challenges to those seeking to address the environmental, climate and social impacts. Some governments have adopted regulations to address a number of impacts, including the European Union (EU) Directive on waste electrical and electronic equipment, the EU Directive on the restriction of certain hazardous substances in electrical and electronic equipment, the EU Regulation on waste shipments, the EU circular economy package, the EU Regulation on conflict minerals, the California electronic waste recycling Act and the US regulation on conflict minerals information disclosure, the so-called Dodd-Franck Act. However, the environmental, climate and social problems often arise in countries without stringent regulation or enforcement. While some policies have extra-jurisdictional effects, overall, regulatory efforts have not reversed substantially existing unsustainable practices (Cuvelier 2017). In the absence of strong public regulation, private codes of conducts and standards have spread (Raj-Reichert 2012). Companies can supplement or even substitute for public welfare and social provisions by adopting political Corporate Social Responsibility programmes (Wickert 2016). Nevertheless, those initiatives have not addressed sufficiently the environmental, climate and social problems. Consequently, the smartphone sector requires further improvements and leaves ample scope for an environmental, climate and social leader to tackle the remaining challenges.

Environmental, climate and social leadership and the smartphone life cycle

This section outlines Liefferink and Wurzel (2017) leadership framework and links it to different follower groups towards which a smartphone company can exert leadership. Leaders explicitly seek to attract followers, while pioneers do not pursue the explicit ambition to attract followers. They implement innovations that go beyond existing ones, an activity that does not necessarily attract followers (Wiering et al. 2018). Liefferink and Wurzel (2017) differentiate four leadership types based on earlier conceptions of environmental leadership (Young 1991, Malnes 1995, Andresen and Agrawala 2002, Skodvin and Andresen 2006, Parker and Karlsson 2010, 2018, Wurzel and Connelly 2011): cognitive, exemplary, entrepreneurial and structural. Each of the types describes a different mechanism that details how a leader can attract followers (see also Wurzel et al. 2019 – this volume).

Cognitive leadership describes the production of ideas, solutions and innovations that shape the perspectives of others. Cognitive leaders provide possible solutions and experiences that inspire followers (Liefferink and Wurzel 2017), including a vision of a desirable future and the ways to reach this vision (Defee et al. 2009). Exemplary leadership consists of the implementation of an idea or innovation in an actor's own practices, policies and organisational structures. Leaders can develop the innovation themselves, in which case exemplary leadership would follow from cognitive leadership, but other sources can also inspire them. Exemplary leadership demonstrates the feasibility of the ambitious idea (Underdal 1994, Malnes 1995, Skodvin and Andresen 2006), which can boost the leader's credibility. Entrepreneurial leadership is the attraction of followers through diplomatic skills, bargaining, persuasion, framing, devising new compromise proposals and creating winning coalitions (Grubb and Gupta 2000, Parker et al. 2012). The leader directly interacts with possible followers, actively tries to convince them, engages in a dialogue and collaborates with them. Structural leadership uses a power position that occurs when a possible follower is dependent on the leader. Structural leaders utilise their dominant position in the international system and the size of their market, finances or other types of power to impose sanctions and provide incentives, pushing actors into certain behaviours (Oberthür and Groen 2015, Liefferink and Wurzel 2017).

Liefferink and Wurzel (2017) originally devised their framework for international negotiations amongst countries. For the smartphone company case, the possible followers are more diverse since the goal of leadership efforts is not achieving ambitious international agreements but rather inducing change in the entire smartphone life cycle. By comparison with companies, a country can separate more neatly the activities of adopting ambitious domestic policy and engaging with external actors. To a larger extent, a smartphone company depends on changes implemented by external supply chain actors to be able to change its own products (Winter and Knemeyer 2013). The divide between internal and external ambition is less clear-cut in this case.

Suppliers of materials and smartphone components are not the only possible targets of a leader's influence. A range of different actor groups can be followers of a smartphone company's environmental, climate and social leadership (see also Torney 2019 – this volume). Achieving more sustainable outcomes requires changes at multiple stages of the smartphone life cycle. Consequently, applying Liefferink and Wurzel's (2017) leadership framework to a broad range of possible followers goes beyond a conventional perspective of sustainable supply chain management. While existing literature generally assumes that companies engage in sustainable supply chain management because of customer, NGO and governmental pressure and in order to avoid reputational risks (Seuring and Müller 2008, Gosling et al. 2016), a social enterprise starts with the aim of achieving a certain environmental, climate or social goal. Instead of responding to pressure, it strives to influence competitors, consumers, end-of-life treatment actors and policymakers, in addition to its suppliers.

These five follower groups are not receptive to the same (combination of) leadership type(s). Followers' specific relation to the leader and their particular characteristics exclude some leadership types and make others more suitable (see Table 1). The broad follower group of suppliers, which includes contract manufacturers, component manufacturers, raw material miners and processors, generally is susceptible to all four leadership types. It is the only one where structural leadership seems a plausible mechanism, exerting influence through the setting of product specifications and codes of conduct. By using audits, leaders can verify suppliers' compliance and punish failures to comply with the loss of the business relationship (van Lakerveld and van Tulder 2017). A small enterprise's leverage is of course limited. A smartphone brand company can exert exemplary leadership through implementing ambitious measures linked to own activities such as greenhouse gas emission reductions and ensuring labour rights. Entrepreneurial leadership can be highly important since it is quite challenging to reach second- and third-tier suppliers through structural leadership. Eliciting a favourable response without using threats or rewards (structural leadership) but rather through persuasion, framing and coalition-building (entrepreneurial leadership) can provide alternative means (Holsti 1983, Allan and Hadden 2017, Murphy-Gregory 2018). Capacity-building to enable an actor to implement changes can help those who, in principle, want to

Table 1. Leadership types and their susceptible follower groups.

Follower groups	Leadership types
Suppliers	**Cognitive**: developing innovative ideas, solutions and expertise **Exemplary**: demonstrating feasibility of higher environmental, climate and social ambition **Entrepreneurial**: actively persuading suppliers, framing the debate, building coalitions and building capacity **Structural**: manipulating utility calculations (through product or process specifications)
Competitors	**Cognitive**: developing innovative ideas, solutions and expertise **Exemplary**: demonstrating feasibility of higher environmental, climate and social ambition **Entrepreneurial**: actively persuading peers, framing the debate and building coalitions
Consumers	**Cognitive**: developing innovative ideas, solutions and expertise **Entrepreneurial**: actively persuading consumers, framing the debate and fostering community building
End-of-life treatment actors	**Cognitive**: developing innovative ideas, solutions and expertise **Exemplary**: demonstrating feasibility of higher environmental, climate and social ambition **Entrepreneurial**: actively persuading recyclers and e-waste collectors, framing the debate, building coalitions and building capacity
Policymakers	**Cognitive**: developing innovative (policy) ideas, solutions and expertise **Exemplary**: demonstrating feasibility of higher environmental, climate and social ambition **Entrepreneurial**: actively persuading policymakers, framing the debate and fostering coalition-building

change but face limitations due to capacity constraints (Biedenkopf *et al.* 2017). Scholars and practitioners have long recognised that working directly with suppliers to train and enable them to comply with ambitious requirements is an important practice (Krause *et al.* 1998, Modi and Mabert 2007, Winter and Knemeyer 2013). Buyers and suppliers can engage in a dialogue that involves the transfer of knowledge, resources and organisational practices to build supplier capacity and enable them to implement innovations (van Lakerveld and van Tulder 2017).

The particularity of a social enterprise is its prioritisation of a societal goal over an economic one. For this reason, other smartphone brand companies are one of their targeted follower groups, principally susceptible to cognitive, exemplary and entrepreneurial leadership. The production of novel ideas, solutions and expertise can influence competitors' practices and products. Exemplary leadership can demonstrate the feasibility of certain practices, which through NGO and public pressure can influence competitors. An entrepreneurial leader's active outreach and coalition-building can influence competitors.

Consumers are an important follower group to encourage suitable and prolonged smartphone use, but they are susceptible to only two of the leadership types. Promoting innovative ideas, solutions and expertise can influence and enable consumers to change their behaviour (cognitive leadership). This includes modular design that enables and encourages consumers to replace individual parts rather than discarding the entire smartphone. Providing adequate information can generate consumer support (Dermont 2018). Entrepreneurial leadership through active persuasion and framing the debate can change consumer behaviour. Building a community in which members collaborate and mutually encourage each other can be an influential tool.

Cognitive and entrepreneurial leadership are the leadership tools towards end-of-life treatment actors. Exemplary leadership can be exerted not through examples of the exact same business practice but through implementing ambitious measures linked to a company's general operation such as reducing greenhouse gas emissions and ensuring labour rights.

Policymakers can be receptive to a smartphone company's cognitive leadership through the provision of innovative policy ideas, solutions and expertise in the policymaking process. Through exemplary leadership, a company can demonstrate that it is technologically and economically feasible to produce devices that are more environmentally, climate and socially friendly. This can be a powerful argument in political debates, rebutting other companies' scepticism and arguments. Actively persuading policymakers, framing the political debate and fostering coalition-building amongst policymakers and stakeholders (entrepreneurial leadership) can influence policy decisions (Carter and Childs 2018, Dupuis and Schweizer 2019, – this volume). Table 1 summarises the susceptibility of each follower group to the leadership types.

One leadership type alone seems unlikely to trigger change. Leadership based on mutually reinforcing types seems more likely to bear success (Underdal 1994, Parker *et al.* 2012, Liefferink and Wurzel 2017, Wiering *et al.* 2018). The four types can amplify each other's effect so that the result is more than the sum of the individual activities. For example, ideas and knowledge (cognitive leadership) but also the adoption of ambitious internal measures (exemplary leadership) need to be brought to the attention of possible followers through diplomatic efforts (entrepreneurial leadership).

Fairphone's environmental, climate and social leadership

This section applies the leadership framework to the social enterprise Fairphone. Entrepreneurial leadership accounts for a substantial share of Fairphone's activities. Since this leadership type is particularly capacity-intensive but essential to achieve its goals, Fairphone took an experimental and gradual approach to expanding and strengthening its leadership. Over the course of its 6 years of operation and with the production of a first and second generation of smartphones, Fairphone has extended and adjusted its leadership activities as a result of learning and experimentation. We base our analysis on an in-depth study of the plethora of information available on the company's website and third-party reporting. Given Fairphone's commitment to transparency, the large number and great detail of blog posts and reports by various members of staff and external experts provides ample information for a thorough mapping. We complement this with third-party information wherever available.

At an overarching level and encompassing all target groups, Fairphone exerts cognitive leadership by developing the idea of a 'fair phone'. Defining this concept is challenging and a continuous process. It comprises all life cycle stages including design, production, distribution, use and end-of-life treatment. No individual aspect might seem revolutionary, but the idea of producing a fair and sustainable phone is innovative and goes beyond other smartphone companies' efforts. Fairphone pursues the goal of designing smartphones that incorporate the ideas of longevity, ownership and fairness (Hebert 2015b) 'to improve the electronics value chain one step at a time' (Fairphone 2016a). The goal goes beyond producing a more environmentally and socially sustainable phone but rather includes compelling broader change beyond Fairphone's own operations. Three Dutch organisations – the foundation *Waag Society*, the NGO *Action Aid Netherlands* and the consultancy *Schrijf-Schrijf* – founded Fairphone in 2010. In 2013, it started selling its first 'fair phone' and launched in 2016 its second generation of smartphones – the *Fairphone 2*.

Fairphone's exemplary leadership still has to catch up with its cognitive leadership. Fairphone has defined the idea and pioneered the production of a fairer smartphone, but it does not produce the most sustainable smartphone imaginable and has not yet put into practice all of its ambitions. Nonetheless, Fairphone has raised the bar and is the company with the best score according to a number of studies. For example, Greenpeace awards the Fairphone 2 ten out of ten possible points for repairability, and a 'B' score – the highest of all brands – in a report on green electronics (Greenpeace 2017a, 2017b, 2017c). It was the first smartphone to receive the Blue Angel certification, a German ecolabel (Ballester Salvà 2016a). In a 2015 comparison between Fairphone and the first smartphone to receive a Swedish Confederation of Professional Employees (TCO) sustainability certification, the Fairphone scored significantly better; nevertheless reports have highlighted the scope for further improvement (Alvord *et al.* 2004, Luke and Chu 2013). Elimination of hazardous substances and reducing its own manufacturing carbon impact are two examples where experts identify scope for improvement (Jardim 2017, Greenpeace 2017a).

Most leadership activities are targeted towards the follower group of suppliers. Fairphone has exerted cognitive and some exemplary leadership by developing and implementing a business model that goes beyond certification and monitoring. It rather strives to engage directly with its supply chain, reaching beyond first-tier suppliers (Ballester Salvà 2015). Fairphone has only exerted limited structural leadership. For the production of the Fairphone 2, the company strove to exert more influence on its supply chain by changing its Chinese contract manufacturer. This was possible because Fairphone had developed into a more attractive production partner for suppliers after its Fairphone 1 success (Hebert 2015c).

Entrepreneurial leadership makes up the largest share of Fairphone's activities towards suppliers. In addition to setting sustainability requirements, the company emphasises the collaborative identification of improvement opportunities and the joint creation of an improvement plan for its suppliers (Lempers 2016). This is of course not an innovation in itself; still, the intensity and emphasis seem to exceed those of other smartphone companies which often focus on first-tier suppliers only (Locke and Samel 2012, Nadvi and Raj-Reichert 2015). Fairphone gradually expanded its leadership by progressively addressing different conflict minerals, including tin, tantalum, gold and tungsten (Gerritsen 2016a).

Fairphone emphasises the importance of relationship building with its supply chain partners to avoid suppliers simply 'ticking boxes' without implementing structural changes. It produces the Fairphone 2 in small batches with a short production cycle to ensure both continuous engagement with the factory and a phased approach to long-term improvement of the suppliers' activities (Ansett 2013). Close collaboration with selected suppliers includes the provision of tools to uncover and address the underlying environmental, climate and social problems (Fairphone 2017a). For Fairphone 1 production, Fairphone paid most attention to first-tier suppliers while, for Fairphone 2 production, it broadened efforts and augmented ambitions. Fairphone has mapped its direct (*first-tier*) and indirect (*second, third and fourth-tier*) suppliers and successively includes suppliers further up the supply chain to establish relationships and improvement programmes (Fairphone 2016b). Fairphone also collaborates with other social enterprises such as Nager-IT, which develops fair cables, building a coalition of social enterprises (Jordan 2013). With the Austrian tungsten smelter Wolfram Bergbau and Hütten AG, Fairphone collaborated to support the mining of conflict-free tungsten from Rwanda (Gerritsen 2016b). Fairphone collaborates with an Eindhoven-based distribution centre to optimise its smartphone packaging (Ballester Salvà 2016b). We can also observe the collaborative nature of Fairphone's approach in the Workers Welfare Fund established together with its supplier Guohong. Both companies contribute $2.5 per produced phone to the fund (Ansett 2014).

Fairphone exerts entrepreneurial leadership towards suppliers not only bilaterally but also through collaboration with other like-minded actors in various initiatives. Together with the Dutch government and other actors, Fairphone participated in the Conflict-Free Tin Initiative, a project seeking to demonstrate that companies can source conflict-free minerals from the Democratic Republic of Congo. Under the Dutch government's brokerage, various supply chain actors collaborated to ensure conflict-free tin sourcing (Conflict-Free Tin Initiative 2017). Another example is Fairphone's effort to ensure conflict-free gold. Jointly with the Austrian printed circuit board manufacturer AT&S, Fairphone mapped the origins of the gold and developed a solution through which it could trace its gold up the supply chain, as the first manufacturer in the sector (Gerritsen 2015). Through cooperation with several of its supply chain stakeholders, Fairphone succeeded in persuading its soldering paste manufacturer to create a separate production line that allows the tracing of tin from mine to phone (Gerritsen 2013b).

Cognitive and especially exemplary leadership form part of Fairphone's strategy towards the follower group of competitors. It is a trailblazer, but we have not yet collected proof of its influence on other smartphone companies. Fairphone engaged in cognitive leadership towards its competitors by developing ideas and solutions for a 'fairer' phone. It established working groups that include participants from civil society, academia and business to develop ideas for aspects such as responsible mining in the Democratic Republic of Congo, working conditions in China and the environmental impact of smartphones (Bleekemolen 2013), which can influence other smartphone companies to include those ideas and solutions in their own practices.

Exemplary leadership by proving that demand exists for a fair phone and by starting a movement towards increased smartphone sector sustainability in order to set an example for other manufacturers is one of Fairphone's explicit goals (see for example Huisken 2015, Fairphone 2017b). Fairphone demonstrated the feasibility of improving the sustainability of smartphone production and has progressively increased the sustainability of its supply chain and product design while selling more than 100,000 smartphones between 2013 and 2016 (Mier 2016). One illustration of its exemplary leadership is Fairphone's development of a repair manual that exceeds any competitor's products repair facilitation (Wiens 2014). These aspects make Fairphone a pioneer. No one so far has collected evidence of these pioneering activities influencing other smartphone companies.

Entrepreneurial leadership forms a lesser part of Fairphone's activities towards competitors. For example, Fairphone participated in the 2014 World Mobile Congress with the aim to bring a socioenvironmental dimension into the technology-focused conference (Xiao 2014). In cooperating with Vodafone and using the Vodafone eco-score, Fairphone makes its phone's environmental impact comparable to that of others while simultaneously providing input into the further development of the eco-score. Fairphone hopes this will inspire other companies (de Jong 2012).

Entrepreneurial leadership dominates Fairphone's approach towards the follower group of consumers. Building a community, dialogue with and empowerment of consumers, for example, by providing online repair tutorials (Fairphone 2017c), is central to Fairphone's consumer strategy. Fairphone promotes 'true ownership' and provides various tips to extend its phones' lifetimes (Tchakaloff 2016). The Fairphone operating system is open source to encourage consumers to use and possibly improve it (Jongenburger 2016). Through transparency in financial aspects, breaking down the costs of a Fairphone (van Abel 2015) and in all its operations along the supply chain, the company aims to include and engage its users and make them part of a community. Fairphone moreover engages in consumer awareness raising through, for example, workshops and a manual on urban mining (Koreniushkina 2015).

Fairphone only indirectly targets the follower group of end-of-life treatment actors through cognitive leadership by strongly investing in smartphone design that enables repair and reuse (Hebert 2015a). By designing a modular phone, Fairphone enables consumers to extend the lifetime of its product and disrupts the rapid replacement cycle characterising large parts of the smartphone industry (van Abel 2016). Compared to conventional smartphone companies, Fairphone designs to a larger extent its products for repairability and refurbishment, and the company provides spare parts to encourage and enable phone repairs (Stoop 2016).

Fairphone mainly targets policymakers in various countries as a follower group through entrepreneurial and cognitive leadership. For example, it collaborated with the Ghanaian Environmental Protection Agency in organising a workshop to explore ways in which the Ghanaian government could address the country's electronic waste problem (de Kluiver 2015). Fairphone provides input to EU consultations (Gerritsen 2014). At a 2013 European Parliament conference on responsible sourcing in conflict regions, its representative shared the company's experiences while at the same time engaging in advocacy for ambitious policy (Gerritsen 2013a). As another illustration, Fairphone's CEO participated in a 2012 OECD forum on due diligence on tin (soldering parts), tantalum (phone capacitors), tungsten (vibrators) and gold (connections). He contributed insights into how companies can develop due diligence processes and seized the networking opportunity with peers (van Abel 2012).

Table 2 highlights the core features of Fairphone's application of the different leadership types. The bulk of activities focuses on suppliers ranging from mineral mines to contract manufacturers, where entrepreneurial

Table 2. Fairphone's application of the different leadership types.

	Cognitive leadership	Exemplary leadership	Entrepreneurial leadership	Structural leadership
Suppliers	Ambitious ideas	Limited	Well developed; bilateral and multipartner collaboration	Limited
Competitors	Ambitious ideas pioneer (not yet a leader)	Proof of concept; pioneer (not yet a leader)	Limited	n/a
Consumers	Ambitious ideas	Limited	Community building	n/a
End-of-life treatment actors	Indirect through product design	Limited	Limited	n/a
Policymakers	Expertise	Limited	Some collaboration	n/a

leadership dominates. It has resulted in a number of changes of practices as a result of entrepreneurial leadership combined with cognitive leadership. Competitors are explicit addressees of Fairphone's leadership efforts, but evidence of followership is not (yet) available. Exemplary leadership is pronounced towards competitors and dominates conjointly with cognitive leadership. Community building and empowerment through entrepreneurial leadership characterise Fairphone's approach towards consumers. Fairphone does not directly address end-of-life treatment actors – only indirectly through the improved smartphone reparability and recyclability. Leadership towards policymakers consists of cognitive leadership in the form of contributing expertise to policy debates but does not appear very pronounced.

The mapping and analysis of Fairphone's leadership activities demonstrates that leadership is better developed and more elaborate towards some follower groups than others and that different follower groups are susceptible to different (combinations of) leadership types. Applying Liefferink and Wurzel (2017) framework can reveal fine-grained insights but benefits from the additional subdivision and separate analysis of the five different follower groups. This highlights that not only the leader's properties and activities but also the possible follower's characteristics and receptiveness play a meaningful role for understanding leadership.

Conclusions

Fairphone originated with the explicit ambition to be a sustainability leader in a complex sector that poses major leadership challenges for a small social enterprise. For this reason, its leadership is gradually expanding and strengthening along with the company's growth and based on experimentation and learning. Achieving a goal such as transforming the smartphone sector requires a step-by-step approach, targeting different follower groups that range from suppliers, competitors and consumers to end-of-life processors and policymakers. In its short period of operation, Fairphone has not always achieved its goals, in part, because of a lack of power and resources (Jongenburger 2014). It has focused more leadership activities on suppliers and consumers than on competitors, end-of-life processors and policymakers. This seems in line with an approach that prioritises core aspects of achieving the overall goal.

Here, we focus on mapping and understanding Fairphone's leadership activities. While we refer to goal achievement and effectiveness in the empirical part, we do not analyse it systematically. Understanding and disentangling the different leadership activities, as this contribution does, can build a solid foundation for a systematic effectiveness analysis, evaluating the extent to which the leadership activities have generated external effects and attracted followers. An analysis of the effectiveness of

Fairphone's environmental, climate and social leadership could be insightful follow-up research to the analysis presented here.

Environmental and climate leadership literature in political science has so far predominantly investigated countries and international negotiations. We have proposed a way in which those leadership concepts can apply to a non-state actor, more specifically to a social enterprise. Our addition of different follower groups to Liefferink and Wurzel's (2017) framework demonstrates that the followership is an important variable in leadership studies (see also Torney 2019 – this volume). In international negotiations, the targets of leadership activities are generally other countries. In the case of a social enterprise, the range of possible followers is more diverse, as we have shown. Not only taking the followers and their susceptibility into account but also differentiating between follower groups with divergent susceptibilities can contribute to better understanding leadership success and failure.

While we have applied the framework to one case in an exploratory manner, future research could apply it to various other types of non-state actors, including conventional companies, to systematically generate comparative insights into the role that non-state actors can play as leaders in sustainable transitions. Fairphone is a likely case for environmental, climate and social leadership since the enterprise's explicit goal is to improve the global smartphone sector's sustainability. Multinationals rather pursue commercial priorities, contrary to social enterprises. Due to their size, they could however generate greater or different impacts on their value chains. Our Fairphone study can provide a reference point for exploring and comparing those aspects.

Acknowledgments

The KU Leuven Research Fund (Bijzonder Onderzoeksfonds KU Leuven) under grant STG/14/039 supported this work. We presented an earlier version at the INOGOV workshop 'Pioneers and Leaders in Polycentric Climate Governance' (PiLePoC) at the University of Hull, September 2016. We thank all workshop participants for their insightful comments and suggestions. We are particularly grateful to Diarmuid Torney, Duncan Liefferink, Rüdiger Wurzel, Hayley Walker and the anonymous reviewers for their help.

Disclosure statement

No potential conflict of interest was reported by the authors.

ORCID

Katja Biedenkopf ⓘ http://orcid.org/0000-0002-2892-3451
Kris Bachus ⓘ http://orcid.org/0000-0002-3150-2068

References

Fairphone, 2016b. *List of suppliers for the Fairphone 2*. [online] Amsterdam: Fairphone. Available from https://www.fairphone.com/wp-content/uploads/2016/10/List-of-Suppliers_Aug2016.pdf [Accessed 10 July 2017].

Allan, J.I. and Hadden, J., 2017. Exploring the framing power of NGOs in global climate politics. *Environmental Politics*, 26 (4), 600–620. doi:10.1080/09644016.2017.1319017.

Alvord, S.H., Brown, L.D., and Letts, C.W., 2004. Social entrepreneurship and societal transformation: an exploratory study. *The Journal of Applied Behavioural Science*, 40 (3), 260–282. doi:10.1177/0021886304266847.

Andresen, S. and Agrawala, S., 2002. Leaders, pushers and laggards in the making of the climate regime. *Global Environmental Change*, 12 (1), 41–51. doi:10.1016/S0959-3780(01)00023-1.

Ansett, S., 2013. *Made with care: social assessment report*. [online] Amsterdam: Fairphone, Available from https://www.fairphone.com/en/2013/12/10/made-with-care-social-assessment-report/ [Accessed 10 July 2017].

Ansett, S., 2014. *Establishing a worker welfare fund with our production partner Guohong*. [online] Amsterdam: Fairphone, Available from: https://www.fairphone.com/en/2014/05/15/establishing-a-worker-welfare-fund-with-our-production-partner-guohong/ [Accessed 10 July 2017].

Ballester Salvà, M., 2015. *Comparing Fairphone's approach to a sustainability label*. [online] Amsterdam: Fairphone, Available from: https://www.fairphone.com/en/2015/07/23/comparing-fairphones-approach-to-a-sustainability-label/ [Accessed 10 July 2017].

Ballester Salvà, M., 2016a. *Fairphone 2 is first smartphone to receive blue angel certification*. [online] Amsterdam: Fairphone, Available from: https://www.fairphone.com/en/2016/10/24/fairphone-2-first-smartphone-receive-blue-angel-certification/ [Accessed 10 July 2017].

Ballester Salvà, M., 2016b. *From the factory to you: packaging the Fairphone 2*. [online] Amsterdam: Fairphone, Available from: https://www.fairphone.com/en/2016/04/22/from-the-factory-to-you-packaging-the-fairphone-2-2/ [Accessed 10 July 2017].

Biedenkopf, K., 2015. Hazardous waste. *In*: P. Pattberg and F. Zelli, eds. *Encyclopedia of global environmental governance and politics*. Cheltenham: Edward Elgar, 380–387.

Biedenkopf, K., Van Eynde, S., and Walker, H., 2017. Policy infusion through capacity-building and project interaction: greenhouse gas emissions trading in China. *Global Environmental Politics*, 17 (3), 91–114. doi:10.1162/GLEP_a_00417.

Bleekemolen, B., 2013. *Our approach to research: forming working groups*. [online] Amsterdam: Fairphone, Available from: https://www.fairphone.com/en/2013/10/24/our-approach-to-research-forming-working-groups/ [Accessed 10 July 2017].

Carter, N. and Childs, M., 2018. Friends of the earth as a policy entrepreneur: 'The Big Ask' campaign for a UK climate change act. *Environmental Politics*, 27 (6), 1–20.

Conflict-Free Tin Initiative, 2017. The Hague: conflict-free tin initiative website [online]. Available from: http://solutions-network.org/site-cfti/ [Accessed 10 July 2017].

Cuvelier, J., 2017. Leaving the beaten track? The EU regulation on conflict minerals. *In*: *Egmont Africa policy brief No. 20*. Brussels: Egmont Institute, 1–7.

Dauvergne, P. and Lister, J., 2013. *Eco-business. a big-brand takeover of sustainability*. Cambridge, MA: MIT Press.
de Jong, M., 2012. *Fairphone becomes a trending topic for 2013*. [online]. Amsterdam: Fairphone, Available from https://www.fairphone.com/en/2012/11/25/trending-topic-for-2013/ [Accessed 10 July 2017].
de Kluiver, J., 2015. *Guest blog: latest news on Ghana e-waste collection program*. [online] Amsterdam: Fairphone, Available from https://www.fairphone.com/en/2015/03/26/latest-news-on-ghana-e-waste-collection-program/ [Accessed 10 July 2017].
Defee, C.C., et al., 2009. Leveraging closed-loop orientation and leadership for environmental sustainability. *Supply Chain Management: An International Journal*, 14 (2), 87–98. doi:10.1108/13598540910941957.
Dermont, C., 2018. Environmental decision-making: the influence of policy information. *Environmental Politics*, online first. doi:10.1080/09644016.2018.1480258.
Dichon, M. and Anderson, A.R., 2009. Social enterprise and effectiveness: a process typology. *Social Enterprise Journal*, 5 (1), 7–29. doi:10.1108/17508610910956381.
Dupuis, J. and Schweizer, R., 2019. Climate pushers or symbolic leaders? the limits to corporate climate leadership by food retailers. *Environmental Politics*, 28 (1).
Egels-Zandén, N., 2016. Not made in China: integration of social sustainability into strategy at Nudie Jeans Co. *Scandinavian Journal of Management*, 32 (1), 45–51. doi:10.1016/j.scaman.2015.12.003.
Egels-Zandén, N., 2017. The role of SMEs in global production networks: a Swedish SME's payment of living wages at its Indian supplier. *Business & Society*, 56 (1), 92–129. doi:10.1177/0007650315575107.
Egels-Zandén, N. and Lindholm, H., 2015. Do codes of conduct improve worker rights in supply chains? a study of Fair Wear Foundation. *Journal of Cleaner Production*, 107, 31–40. doi:10.1016/j.jclepro.2014.08.096.
Ercan, M., et al., 2016. Life cycle assessment of a smartphone. 4th International Conference on ICT for Sustainability (ICT4S 2016), Amsterdam.
Fairphone, 2016a. *Fairphone website*. [online]. Amsterdam: Fairphone. Available from http://www.fairphone.com/ [Accessed 12 September 2016].
Fairphone, 2017a. *Good working conditions*. [online]. Amsterdam: Fairphone. Available from: https://www.fairphone.com/en/our-goals/social-work-values/ [Accessed 10 July 2017].
Fairphone, 2017b. *Our goals*. [online]. Amsterdam: Fairphone. Available from: https://www.fairphone.com/en/our-goals/ [Accessed 10 July 2017].
Fairphone, 2017c. *Welcome to Fairphone support*. [online]. Amsterdam: Fairphone. Available from: https://fairphone.zendesk.com/hc/en-us [Accessed 10 July 2017].
Gereffi, G., Humphrey, J., and Sturgeon, T., 2005. The governance of global value chains. *Review of International Political Economy*, 12 (1), 78–104. doi:10.1080/09692290500049805.
Gerritsen, L., 2013a. *Fairphone at European Parliament*. [online] Amsterdam: Fairphone, Available from: https://www.fairphone.com/en/2013/06/13/fairphone-at-european-parliament/ [Accessed 10 July 2017].
Gerritsen, L., 2013b. *Tin and tantalum road trip*. [online] Amsterdam: Fairphone, Available from: https://www.fairphone.com/en/2013/11/08/tin-and-tantalum-road-trip/ [Accessed 10 July 2017].
Gerritsen, L., 2014. *Conflict-free mineral legislation in the US and EU*. [online] Amsterdam: Fairphone, Available from: https://www.fairphone.com/en/2014/04/16/conflict-free-mineral-legislation-in-the-us-and-eu/ [Accessed 10 July 2017].

Gerritsen, L., 2015. *Digging deeper into the gold supply chain with our partner AT&S.* [online] Amsterdam: Fairphone, Available from: https://www.fairphone.com/en/2015/11/11/digging-deeper-into-the-gold-supply-chain-with-our-partner-ats/ [Accessed 10 July 2017].

Gerritsen, L., 2016a. *Fairphone 2 good vibrations with conflict-free tungsten.* [online] Amsterdam: Fairphone, Available from: https://www.fairphone.com/en/2016/06/20/fairphone-2-good-vibrations-with-conflict-free-tungsten-2/ [Accessed 10 July 2017].

Gerritsen, L., 2016b. *Supporting conflict-free tungsten in Rwanda.* [online] Amsterdam: Fairphone, Available from: https://www.fairphone.com/en/2016/01/13/supporting-conflict-free-tungsten-in-rwanda/ [Accessed 10 July 2017].

Gosling, J., et al., 2016. The role of supply chain leadership in the learning of sustainable practice: toward an integrated framework. *Journal of Cleaner Production*, 137, 1458–1469. doi:10.1016/j.jclepro.2014.10.029.

Greenpeace, 2008. *Chemical contamination at e-waste recycling and disposal sites in Accra and Korforidua, Ghana.* Amsterdam: Greenpeace.

Greenpeace, 2017a. *Guide to greener electronics. 2017 company report card.* Washington, DC: Greenpeace.

Greenpeace, 2017b. *How repairable is your mobile device? a product guide to best-selling smartphones, tablets and laptops.* Taipei: Greenpeace East Asia.

Greenpeace, 2017c. *Rethink IT.* [online]. Amsterdam: Greenpeace International. Available from: https://www.rethink-it.org/en/ [Accessed 10 July 2017].

Grubb, M. and Gupta, J., 2000. Leadership: theory and methodology. In: J. Gupta and M. Grubb, eds. *Climate change and European leadership: a sustainable role for Europe?* Dordrecht: Kluwer Academic Publishers, 15–24.

Hebert, O., 2015a. *The architecture of the Fairphone 2: designing a competitive device that embodies our values.* [online] Amsterdam: Fairphone, Available from https://www.fairphone.com/en/2015/06/16/the-architecture-of-the-fairphone-2-designing-a-competitive-device-that-embodies-our-values/ [Accessed 10 July 2017].

Hebert, O., 2015b. *Designing the next Fairphone from the inside out.* [online] Amsterdam: Fairphone, Available from: https://www.fairphone.com/en/2015/05/13/designing-the-next-fairphone-from-the-inside-out/ [Accessed 10 July 2017].

Hebert, O., 2015c. *The path to finding our new production partner: hi-P.* [online] Amsterdam: Fairphone, Available from: https://www.fairphone.com/en/2015/02/19/the-path-to-finding-our-new-production-partner-hi-p/ [Accessed 10 July 2017].

Holsti, K.J., 1983. *International politics. A framework for analysis.* London: Prentice Hall.

Huisken, M., 2015. *Guest blog. iFixit on Fairphone 2: the first truly smart smartphone.* [online] Amsterdam: Fairphone, Available from: https://www.fairphone.com/en/2015/11/18/guest-blog-ifixit-on-fairphone-2-the-first-truly-smart-smartphone/ [Accessed 10 July 2017].

Jardim, E., 2017. *From smart to senseless: the global impact of 10 years of smartphone.* Washington, DC: Greenpeace.

Jongenburger, K., 2014. *Our approach to software and ongoing support for the first Fairphones.* [online] Amsterdam: Fairphone, Available from https://www.fairphone.com/en/2014/12/09/our-approach-to-software-and-ongoing-support-for-the-first-fairphones/ [Accessed 10 July 2017].

Jongenburger, K., 2016. *Releasing the Fairphone 2 open operating system.* [online] Amsterdam: Fairphone, Available from: https://www.fairphone.com/en/2016/04/28/releasing-the-fairphone-2-open-operating-system-2/ [Accessed 10 July 2017].

Jordan, S., 2013. *Guest blog: fair cables for fair electronics.* [online] Amsterdam: Fairphone, Available from: https://www.fairphone.com/en/2013/07/19/guest-blog-fair-cables-for-fair-electronics/ [Accessed 10 July 2017].

Koreniushkina, D., 2015. *Host an urban mining workshop.* [online] Amsterdam: Fairphone, Available from: https://www.fairphone.com/en/2015/03/02/host-an-urban-mining-workshop/ [Accessed 10 July 2017].

Krause, D.R., Handfield, R.B., and Scannell, T.V., 1998. An empirical investigation of supplier development: reactive and strategic processes. *Journal of Operations Management*, 17 (1), 39–58. doi:10.1016/S0272-6963(98)00030-8.

Lempers, M., 2016. *Progress and challenges to improving working conditions with our Fairphone 2 manufacturer.* [online] Amsterdam: Fairphone, Available from: https://www.fairphone.com/en/2016/08/18/progress-challenges-improving-working-conditions-fairphone-2-manufacturer/ [Accessed 10 July 2017].

Liefferink, D. and Wurzel, R.K.W., 2017. Environmental leaders and pioneers: agents of change? *Journal of European Public Policy*, 24 (7), 951–968. doi:10.1080/13501763.2016.1161657.

Locke, R.M., Amengual, M., and Mangla, A., 2009. Virtue out of necessity? Compliance, commitment, and the improvement of labor conditions in global supply chains. *Politics & Society*, 37 (3), 319–351. doi:10.1177/0032329209338922.

Locke, R.M. and Samel, H.M., 2012. Looking in the wrong place? Labor standards and upstream business practices in the global electronics industry. *MIT Political Science Department Working Paper No. 2012-18*, Boston.

Luke, B. and Chu, V., 2013. Social enterprise versus social entrepreneurship: an examination of the 'why' and 'how' in pursuing social change. *International Small Business Journal*, 31 (7), 764–784. doi:10.1177/0266242612462598.

Lund-Thomsen, P. and Lindgreen, A., 2014. Corporate social responsibility in global value chains: where are we now and where are we going? *Journal of Business Ethics*, 123 (1), 11–22. doi:10.1007/s10551-013-1796-x.

Malnes, R., 1995. 'Leader' and 'entrepreneur' in international negotiations: a conceptual analysis. *European Journal of International Relations*, 1 (1), 87–112. doi:10.1177/1354066195001001005.

Mier, J., 2016. *We've reached 100,000 Fairphone owners!* [online] Amsterdam: Fairphone, Available from: https://www.fairphone.com/en/2016/05/26/100000-fairphone-owners/ [Accessed 10 July 2017].

Modi, S.B. and Mabert, V.A., 2007. Supplier development: improving supplier performance through knowledge transfer. *Journal of Operations Management*, 25 (1), 42–64. doi:10.1016/j.jom.2006.02.001.

Moran, C.J., et al., 2014. Sustainability in mining, minerals and energy: new processes, pathways and human interactions for a cautiously optimistic future. *Journal of Cleaner Production*, 84, 1–15. doi:10.1016/j.jclepro.2014.09.016.

Murphy-Gregory, H., 2018. Governance via persuasion: environmental NGOs and the social licence to operate. *Environmental Politics*, 27 (2), 320–340. doi:10.1080/09644016.2017.1373429.

Nadvi, K. and Raj-Reichert, G., 2015. Governing health and safety at lower tiers of the computer industry global value chain. *Regulation & Governance*, 9 (3), 243–258. doi:10.1111/rego.12079.

Newell, P., 2001. Managing multinationals: the governance of investment for the environment. *Journal of International Development*, 13 (7), 907–919. doi:10.1002/(ISSN)1099-1328.

Oberthür, S. and Groen, L., 2015. The effectiveness dimension of the EU's performance in international institutions: towards a more comprehensive assessment framework. *Journal of Common Market Studies*, 53 (6), 1319–1335. doi:10.1111/jcms.12279.

Parker, C.F., et al. 2012. Fragmented climate change leadership: making sense of the ambiguous outcome of cop-15. *Environmental Politics*, 21 (2), 268–286. doi:10.1080/09644016.2012.651903.

Parker, C.F. and Karlsson, C., 2010. Climate change and the European Union's leadership moment: an inconvenient truth? *Journal of Common Market Studies*, 48 (4), 923–943. doi:10.1111/j.1468-5965.2010.02080.x.

Parker, C.F. and Karlsson, C., 2018. The UN climate change negotiations and the role of the United States: assessing American leadership from Copenhagen to Paris. *Environmental Politics*, 27 (3), 519–540. doi:10.1080/09644016.2018.1442388.

Peattie, K. and Morley, A., 2008. Eight paradoxes of the social enterprise research agenda. *Social Enterprise Journal*, 4 (2), 91–107. doi:10.1108/17508610810901995.

Raj-Reichert, G., 2012. *Governance in global production networks: managing environmental health risks in the personal computer production chain*. Doctoral Thesis submitted to the University of Manchester.

Raj-Reichert, G., 2015. Exercising power over labour governance in the electronics industry. *Geoforum*, 67, 89–92. doi:10.1016/j.geoforum.2015.10.013.

Sarath, P., et al., 2015. Mobile phone waste management and recycling: views and trends. *Waste Management*, 46, 536–545. doi:10.1016/j.wasman.2015.09.013.

Seuring, S. and Müller, M., 2008. From a literature review to a conceptual framework for sustainable supply chain management. *Journal of Cleaner Production*, 16 (15), 1699–1710. doi:10.1016/j.jclepro.2008.04.020.

Skodvin, T. and Andresen, S., 2006. Leadership revisited. *Global Environmental Politics*, 6 (3), 13–27. doi:10.1162/glep.2006.6.3.13.

Stoop, M., 2016. *How we are fixing the spare parts supply chain so you can repair your Fairphone*. [online] Amsterdam: Fairphone, Available from: https://www.fairphone.com/en/2016/03/31/how-we-are-fixing-the-spare-parts-supply-chain-so-you-can-repair-your-fairphone/ [Accessed 10 July 2017].

Tchakaloff, B., 2016. *Boosting the life of your Fairphone battery*. [online] Amsterdam: Fairphone, Available from: https://www.fairphone.com/en/2016/03/23/boosting-the-life-of-your-fairphone-battery/ [Accessed 10 July 2017].

Torney, D., 2019. Follow the leader? Towards a conceptualization of climate 'Followership'. *Environmental Politics*, 28, 1.

Underdal, A., 1994. Leadership theory: rediscovering the arts of management. In: W. Zartman, ed. *International multilateral negotiation. approaches to the management of complexity*. San Francisco: Jossey-Bass, 178-197.

UNECA, 2011. *Minerals and Africa's development: the International Study Group report on Africa's mineral regimes*. Addis Ababa: Economic Commission for Africa.

van Abel, B., 2012. *Just another boring abstract high-level meeting? OECD conflict minerals*. [online] Amsterdam: Fairphone, Available from: https://www.fairphone.com/en/2012/11/30/oecd-conflict-minerals-just-another-boring-abstract-high-level-meeting/ [Accessed 10 July 2017].

van Abel, B., 2013. *Fairphone's crowdfunded model & social impact*. [online] Amsterdam: Fairphone, Available from: https://www.fairphone.com/en/2013/05/24/crowdfunded-model/ [Accessed 10 July 2017].

van Abel, B., 2015. *Cost breakdown of the Fairphone 2*. [online] Amsterdam: Fairphone, Available from https://www.fairphone.com/en/2015/09/09/cost-break down-of-the-fairphone-2/ [Accessed 10 July 2017].

van Abel, B., 2016. *Fairphone 2 gets a new case design in four colors*. [online] Amsterdam: Fairphone, Available from: https://www.fairphone.com/en/2016/10/18/fairphone-2-gets-new-case-design-four-colors/ [Accessed 10 July 2017].

van Lakerveld, A. and van Tulder, R., 2017. Managing the transition to sustainable supply chain management practices: evidence from Dutch leader firms in Sub-Saharan Africa. *Review of Social Economy*, 75 (3), 255–279. doi:10.1080/00346764.2017.1286033.

Veale, E., 2012. Is there blood on your hands-free device: examining legislative approaches to the conflict minerals problem in the Democratic Republic of Congo. *Cardozo Journal of International and Competition Law*, 21, 503–544.

Wickert, C., 2016. 'Political' corporate social responsibility in small- and medium-sized enterprises: a conceptual framework. *Business & Society*, 55 (6), 792–824. doi:10.1177/0007650314537021.

Wiens, K., 2014. *Guest blog: iFixit and Fairphone repair guides*. [online] Amsterdam: Fairphone, Available from: https://www.fairphone.com/en/2014/04/03/ifixit-guest-blog/ [Accessed 10 July 2017].

Wiering, M., Liefferink, D., and Beijen, B., 2018. The internal and external face of Dutch environmental policy: a case of fading environmental leadership? *Environmental Science and Policy*, 81, 18–25. doi:10.1016/j.envsci.2017.12.002.

Winter, M. and Knemeyer, A.M., 2013. Exploring the integration of sustainability and supply chain management. current state and opportunities for future inquiry. *International Journal of Physical Distribution and Logistics Management*, 43 (1), 18–38. doi:10.1108/09600031311293237.

Wurzel, R. and Connelly, J., 2011. *The European Union as a leader in international climate change politics*. Abingdon: Routledge.

Wurzel, R.K., Liefferink, D., and Torney, D., 2019. Pioneers, leaders and followers in multilevel and polycentric climate governance. *Environmental Politics*, 28 (1).

Xiao, Y., 2014. *Mobile world congress 2014: product trend report*. [online] Amsterdam: Fairphone, Available from: https://www.fairphone.com/en/2014/03/07/mobile-world-congress-2014-product-trend-report/ [Accessed 10 July 2017].

Young, O.R., 1991. Political leadership and regime formation: on the development of institutions in international society. *International Organization*, 45 (3), 281–308. doi:10.1017/S0020818300033117.

Climate pushers or symbolic leaders? The limits to corporate climate leadership by food retailers

Johann Dupuis and Remi Schweizer

ABSTRACT
Corporate climate leadership and its relationship with state regulations are discussed. First, a typology defining corporate climate leadership is introduced and distinguished from the other strategic behaviours corporations may adopt in response to climate change. A conceptual framework to explore the mechanisms enabling corporate climate leadership within a given policy system is then presented. This framework is applied to two big Swiss food retailers, considered as typical of corporate climate leaders, firms that showed an early interest in climate protection, as a result of ecological values, third actors' lobbying and particular market incentives. Most importantly, the two companies were set in motion by a regulatory framework that featured stringent policy goals associated with flexible instruments and economic sanctions. The importance of these findings for understanding the role of corporate leadership in polycentric climate governance is discussed.

Introduction

It is commonplace to emphasise that the state is but one of many actors active in polycentric climate governance and to underline the increasing role played by private actors, transnational non-governmental organisations (NGOs) and major corporations (Andonova *et al.* 2009). The 2015 Paris agreement relies more explicitly than ever on businesses to act as climate champions in order to reach the UNFCCC's ultimate objectives (Hale 2016). Fostering climate initiatives 'from the bottom-up' through flexible regulations that enable innovations from the private sector is a central dimension of the polycentric climate governance in which much hope has been placed (Jordan *et al.* 2015). Here, we explore the mechanisms and regulatory conditions driving firms to assume the much-desired role of *climate leaders* (Wurzel *et al.* 2019 – this volume).

Firms can have an inherent interest in voluntarily deploying climate mitigation measures (Lyon and Maxwell 2008). Indeed, we can associate corporate investments in climate protection with increasing efficiency in resource uses and lower production costs, which allows for improved productivity. In many instances, the expected gains of climate measures outweigh their costs (Matsumura *et al.* 2014).

The persistence of various market failures and information asymmetries that hamper investment in environmental optimisation explains the fact that not all firms are, however, perched on the frontiers of environmental innovation (Lanoie *et al.* 2011). According to the 'Porter hypothesis' (Porter and Van der Linde 1995), the state's main role is to alleviate these inefficiencies through flexible and market-friendly policy instruments, providing information and price incentives that enable firms to efficiently reach the best possible outcome in terms of environmental performance (Jaffe and Palmer 1997). In doing so, these flexible climate policies could foster the emergence of corporate climate leadership.

Scholars have criticised the Porter hypothesis on two accounts. Classical economists contend that the benefits of abatement measures always offset the costs, nor do they agree that state intervention can adequately correct market failures (Palmer *et al.* 1995). By contrast, public policy analysts have questioned the effectiveness of voluntary measures by corporate actors. Indeed, these are often limited to picking the low-hanging fruit, in the sense that only measures with the best cost-efficiency ratio gain traction (Koehler 2007, Pizer *et al.* 2011). More critical voices further analyse corporate responses to climate change as being nothing more than a production of narratives and myths that are used to consolidate current business models and to divert from more radical changes (Wright and Nyberg 2014). In a nutshell, both the willingness of firms to act as climate champions and the capacity of state regulations to foster corporate leadership may face challenges.

The relationship between state regulation and corporate responses is complex. We believe, along with Darnall (2009), that corporate actors do not constitute a homogenous group that uniformly responds to environmental regulations, be they flexible or not. Just as states differ in their attitudes and capacity to act on climate change, some firms might financially benefit and push for climate regulations, while others might not. To understand how corporate climate leadership emerges, there is both a need to discriminate between the variety of corporate behaviours towards climate change and to analyse the role state regulations play in the mechanisms driving corporate leaders to act as such.

We explore the relationship between corporate climate leadership and state regulations, focusing on the mid-range causal mechanisms driving climate leaders in policy contexts characterised by flexibility – thus allowing us to build on the Porter hypothesis. Our contribution first discusses corporate climate leadership based on the Liefferink and Wurzel (2017) typology, which forms the analytical starting point of this Volume. We adjust this typology, previously only applied to states, in order to classify the diversity of behaviours firms may adopt in relation to climate change. Because the typology – like any other ideal-typical construct – has mostly descriptive ambitions, we complement it with an analytical framework to explore the pathways for the emergence of corporate climate leadership. We demonstrate the heuristic value of our conceptual tools through two case studies from a highly strategic sector for climate policies: food retailing. We compare the climate strategies of the two biggest Swiss food retailers (Coop and Migros), which we expect to represent typical cases of corporate climate leaders in a context of flexible regulations. Our results allow for discussion of the qualities and limits of corporate climate leadership, as well as the merits of the Porter hypothesis.

Theories

A typology for analysing corporate climate strategies

Liefferink and Wurzel (2017) propose a two-dimensional matrix in order to qualify states' environmental policies. They distinguish states according to their internal 'face' – the environmental ambitions of domestic policies – and their external 'face' – the environmental ambitions displayed in their foreign policy. *Laggards* have low internal and external merits; *pioneers* are ahead of the pack regarding their domestic policies, but do not pay much attention to attracting followers in international arenas; *symbolic leaders* are demonstrative in international fora, but are not consequential in their domestic policies; *pushers* take the lead domestically, and actively lobby other states to follow their example. Within this frame of reference, *climate leadership* encompasses both pusher and pioneer behaviours.

Scholarly discussion has not so far assessed the relevance of this typology for corporate actors. From an organisation theory perspective, states and firms undeniably are governance structures of a distinct nature (Scott and Meyer 1994). There is, however, nothing that disqualifies the application of the Liefferink and Wurzel framework to corporate actors (see also Wurzel et al. 2019 – this volume). Corporate climate strategies can also have an internal and an external face. The internal climate strategies of businesses can have more or less stringent objectives, be innovative to varying extents, and greatly contribute to lessening the pressure on the climate system in the case of mega-corporations such as Walmart (Lubin and Esty 2010, Stroup and Wong 2018). The internal face here refers to the various measures to reduce the GHG emissions resulting from production processes, transportation and product sales. At the same time, firms also possess an external face. Corporate actors are generally prompt to communicate around their environmental efforts. They may engineer discourses and promote their image of climate champions as part of their marketing.

However, the rationale behind firms' and states' action differs significantly. Indeed, states will generally avoid acting unilaterally on climate change, and try to commit other nations to similar climate protection goals. This is because states often perceive climate policies as costly and as a hurdle to economic competitiveness (Nordhaus 2015). With regards to states, climate policy bears the characteristics of a public good – non-excludability and non-subtractability. This results in a tendency either to act as a free rider, or to attempt to share the cost of climate protection by coordinating with other nations. Thus, we can expect most states willing to take leadership to adopt a pusher rather than a pioneer behaviour.

Firms, by contrast, can use various ways to capture, privatise and exclude others from the benefits of their climate strategies. This is the case when, for instance, a firm succeeds in appearing as the unique climate champion on the market in order to attract the biggest share of 'green customers'. Provided that a limited but growing demand for climate-friendly products exists, firms will attempt to harness their climate efforts as a marketing tool to distance competitors and to grasp the so-called eco-premiums (Lubin and Esty 2010) – market share and profit margins related to green strategies. In other words, with respect to firms and unlike states, climate protection can have the attributes of a private good. Accordingly, we can expect firms willing to act as climate leaders to act as pioneers rather than as pushers.

Whatever the true merits of their internal climate strategy, firms have an inherent interest to exhibit a strong external face and to elaborate a well-designed communication. A structural incentive to act as a *greenwasher* (Lyon and Montgomery 2015) in order to maximize eco-premiums exists among corporate actors. A rational firm would not spend a single penny on implementing measures to protect the climate without attempting to reap the potential benefits in terms of image. Acting any differently would approximate a purely altruistic behaviour by the firm (Lantos 2002, Altman 2005). Strategic renunciation of communicating on climate efforts can sometimes exist, however. Among the many businesses that opposed climate regulations in the United States during the 1990s, some had already taken measures to reduce their GHG emissions while publicly denying the importance of climate change mitigation (Vormedal 2011). Notwithstanding such exceptions, the incentives for all corporations to communicate as if they were climate leaders, regardless of the merits of their actions, make it difficult to differentiate between symbolic leaders, pioneers or pushers.

In order to adapt the Liefferink and Wurzel typology to corporate actors, we suggest disaggregating the 'external face' in two dimensions: a *discursive* 'face', referring to the degree to which corporate actors communicate their actions to consumers; and a *political 'face'* referring to the degree to which corporate actors lobby the state for strengthening climate regulations. Indeed, if the vast majority of corporations market their climate efforts, only a part will concretely engage and support the development of climate change regulations. For instance, firms may be confident enough in the superiority of their CO_2 abatement techniques or in their commercial strategy to view a reinforcement of climate policies as providing them a comparative advantage. The chemical firm Dupont illustrates well this attitude: convinced that the CFC technology was becoming obsolete, Dupont pushed for a new regulation promoting an alternative technology (HCFCs) it had developed, contributing to the success of the Montreal Protocol (Maxwell and Briscoe 1997).

With such a political dimension, the typology allows to distinguish between eight possible categories of firms (see Figure 1).

- *Laggards* have no specific strategies for mitigating GHG emissions, nor do they communicate around the issue; they are either passive or opposed to climate policy;
- *free riders* communicate extensively around climate change and the necessity to take action but, at the same time, do not pursue high ambitions in their internal strategy or in political arenas;
- *symbolic leaders* are proactive in political debates and very openly communicate around the urgency of climate issues without, however,

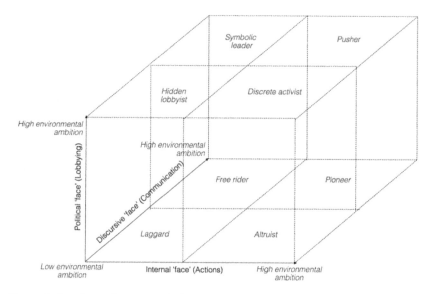

Figure 1. Typology of corporate behaviours on climate change.

translating these ambitions in their own corporate strategies. This type of firm approximates best the figure of the *greenwasher*;
- *hidden lobbyists* do not concretely act or communicate around climate change, but covertly lobby to adopt stronger environmental policies;
- *altruists* are ahead of the pack regarding their corporate strategies, but do not market their climate efforts nor attempt to influence regulations;
- *pioneers* are ahead of the pack regarding their concrete actions and actively communicate about them, but do not engage politically in favour of climate policies;
- *pushers* are innovative and ambitious in terms of internal actions, intensively communicate about them, and actively lobby for the adoption of stronger environmental regulations;
- finally, *discrete activists* are altruistic firms that politically push for stronger environmental regulations and have ambitious internal strategies without communicating about their actions.

These ideal-types constitute theoretical constructs that are more or less likely to exist in reality. We argue that this typology adequately captures the variety of behaviours firms might adopt with regards to climate change. The four types of corporations that have ambitious internal strategies (altruists, pioneers, pushers, discrete activists) can be defined as representing forms of climate leadership. The questions we will further address pertain to how and why private actors act as climate leaders in polycentric settings characterised by flexible climate policies.

A framework to explore corporate climate leadership

Our typology allows for the analysis and classification of firms based on their behaviour towards climate change; however, it does not provide insights on the drivers behind the adoption of a given behaviour.

In order to fill this gap, we propose to complement it with an exploratory framework mainly inspired by public policy theories (Sabatier and Weible 2014) and policy implementation research (Hill and Hupe 2014). Whereas it is true that corporate behaviour is generally a topic of economics and business studies (Lyon and Maxwell 2008), modern policy studies have developed interesting approaches that examine how firms respond or anticipate regulatory stimulus. These studies include factors such as the policy design, ideational orientations, networks and public–private interactions in their models of explanations (Koski and May 2006, Darnall 2009).

Our exploratory framework revolves around three principal dimensions (Figure 2). First, it starts from the premise that firms' actions always take place within a *regulatory framework*. Public regulations can stimulate or impede environmental innovations by firms, depending on their design

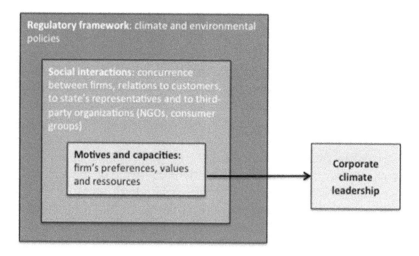

Figure 2. Exploratory framework on corporate climate leadership.

(Porter and Van der Linde 1995). If the Porter hypothesis posits a relation between flexible regulations and climate leadership, we argue that this relationship might not be direct, but mediated through intermediary factors. Our framework therefore aims to identifying the mid-range causal mechanisms that link regulations to firms' responses.

Second, firms' climate strategies are embedded in a web of *social interactions*. Firms constantly adapt to the behaviours of others within complex webs of power (Fligstein, Neil 2001). The bulk of policy studies have traditionally focused on the bidirectional interactions between public and private agents. However, it is increasingly recognised in the literature that decision networks have become more complex, giving more space to non-governmental actors (Rhodes 2007). Economic competition between firms is notably a pattern of interaction that may foster environmental innovation (Nesta *et al.* 2014). Corporate actors are also in constant interaction with environmental NGOs, among other social groups, who pressure them to adopt more or less stringent climate strategies. In the case of food retailers, the exposure to consumer choices and trends may also be a source of influence (Yalabik and Fairchild 2011).

Third, this framework assumes that climate leadership represents a *strategic choice* for corporate actors. Building on actor-centred institutionalism (Scharpf 1997), we argue that corporate choices are strategic because they imply intentional conduct. Firms evaluate, select and constantly adapt the course of their action to the regulatory framework and to the behaviours of other actors (Schweizer 2015). Businesses act in ways they believe to be means to their goals (Korpi 1985). This does not mean these goals are

purely rational and made under perfect cognition, but that climate (in)
action is always motivated. Even where the regulatory framework constrains choices, corporate actors always dispose of some leeway to define
their environmental strategies and have *motives and capacities* to do so.
Firms' motivations either result from their perceived self-interest or their
belief system (Sabatier and Weible 2014). This is the reason why changes of
single individuals in the composition of the company board can influence
tremendously on the values of the firm as a social organisation. Firms'
decisions are also directly related to capacities and resources that can either
represent opportunities for or constraints on action (Knoepfel *et al.* 2011).
A lack of financial resources can make it difficult for small or medium-size
firms to invest in climate innovations, as they often consist of risky investments with uncertain chances of return.

Methods

Our method relies on comparative case studies (Seawright and Gerring
2008). In terms of general heuristics, we follow a procedure akin to the
one Elinor Ostrom and colleagues developed for analysing action situations
(Poteete *et al.* 2010). We use the framework proposed in Figure 2 as an
entry point in empirical case study analysis to identify the pathways to
corporate climate leadership.

Case studies context: the global food sector and the climate

The food sector is a major topic in the climate literature and represents an
excellent field to study climate leadership. Climate change is projected to
undermine food security globally. However, the food industry represents a
major source of global greenhouse gases (GHG). Roughly one fifth of the
overall world GHG emissions in 2010 originated in the food sector (Olivier
et al. 2014).

One of the central levies to achieve systemic GHG abatement in the food
sector might be in the hands of food retailers. In most countries, the advent
of contract farming has contributed to a shift of power from food producers
or manufacturing industries to large retailing companies (Burch and
Lawrence 2005). Big food retailers occupy a pivotal position in which
they can control and monitor the entire supply chain, and impose their
own environmental standards (Vandenbergh 2007). Food retailers influence
not only production patterns (through price and standards setting), but also
consumption habits. By selecting the displayed products, retailers predetermine the carbon footprint of consumers' food basket. Thus, in the food
sector, retailers have become centres of power alternative to the state
(McMichael 2009).

Case selection: Swiss food retailers as typical cases of climate leadership

In order to explore the mechanisms linking corporate climate leadership to flexible regulations, we used the 'typical case' selection strategy as described by Seawright and Gerring (2008). A typical case is perfectly representative of the relationship predicted by a given theory. It allows the researcher to further explore the causal mechanisms behind the predicted relationship in order to complement the theory. According to Seawright and Gerring (2008), a single case study suffices to carry a typical case research strategy. However, to consolidate our findings, we compare two typical cases, which exhibit the exact same flexible regulations, but a slight variation on the dependent variable. This comparative setting corresponds to a 'most similar research design', which is best suited to explore the alternative factors that may have influenced the outcome (Anckar 2008).

Following this logic, we selected the two leaders of the food-retailing sector in Switzerland: Coop and Migros. A benchmark evaluated these two companies to be the most sustainable food retailers in the world (far ahead of Walmart, Carrefour or Tesco) respectively in 2011 and 2014 (Oekom Research AG 2018). Moreover, the Federal Office of the Environment of Switzerland publicly praised Coop and Migros as models of exemplary corporations with regards to climate protection (OFEV 2015). Swiss consumers' willingness to pay for organic products is the highest in the world, as measured by per capita consumption[1], which we can consider as a token of the incentives faced by the two firms to behave as climate leaders.

Swiss climate policy mostly relies on a mix of voluntary programs with businesses in association with market-based instruments, which we consider typical of the flexible policies that should enable climate leadership (Thalmann et al. 2004). At the institutional level, both the principle of the subsidiarity of state intervention and the principle of cooperation with the private sector that are enshrined in public laws create a general setting, which should enable voluntary action and climate innovation by firms (Dupuis and Knoepfel 2015).

Furthermore, selecting two typical cases that evolve in the same regulatory context builds an adequate comparative setting to study whether variations in the two firms' social interactions, motives and capacities (variables of interests) may explain differences in terms of climate leadership. However, limited generalisability is the downside of this research design. We cannot claim that the observed causal patterns would hold in contexts characterised by very different policy frameworks.

Data sources and analysis

We base our study on 10 semi-structured interviews with firms' representatives, public officials and NGOs, one focus group, and content analysis of the grey literature, particularly the numerous reports by the two companies on their sustainable and environmental strategies. We used this material, first, to analyse the climate strategies of the companies based on the three dimensions of our typology, and, second, to identify the mechanisms explaining the adoption or non-adoption of CO_2 reduction measures. The diverse sources were triangulated in order to retrace courses of action and to identify the major drivers/limits to the companies' climate strategies between 1990 and 2015. Based on our empirical material, we wrote a research report draft, which our interviewees, the focus group and the Swiss Federal Office of the Environment validated.

In order to evaluate the ambitions of the firms' internal climate actions, we employed the Greenhouse Gas (GHG) Protocol methodology developed by the World Resources Institutes (WRI) and the World Business Council on Sustainable Development (WBCSD). The GHG Protocol distinguishes between different scopes of emissions. Scope 1 refers to direct emissions related to the main economic activities; Scope 2 deals with the indirect emissions related to the consumption of electricity; Scope 3 includes all upstream and downstream emissions, in particular those embedded in the food (agricultural production) or the transportation and elimination of sold products. Although extremely important in quantity, this third level is generally that most rarely addressed in the literature (Downie and Stubbs 2013). Evaluating the extent to which a firm includes these three scopes provides a way to assess the internal ambition of its climate strategy.

Cases studies: Coop and Migros' climate leadership

Coop and Migros: presenting the firms

Coop and Migros have dominated the Swiss retail food market for decades. Despite the arrival of Aldi and Lidl in the 2000s, they have maintained their positions with a combined share of approximately two-thirds of the market.

Coop (created 1890) is currently number 2. The organisation of the firm is highly vertical, giving extended power to the central direction in Basel. The group is also active in food industries and on the wholesale market. Migros (founded 1925) is the leader in the Swiss retailing sector since 1967. Today, the group counts 10 regional cooperatives and remains rather decentralised in its organization. It is, like Coop, highly integrated and active in food production industries, owning and operating many well-known national food brands.

Coop and Migros initiated environmental initiatives early compared to other Swiss companies. In the 1970s, they implemented their first energy efficiency programs, and created their own organic brands in 1992 (Naturaplan, Coop) and 1995 (Migros Bio). They are undisputed leaders in the organic sector with more than three-quarters of the market (Coop alone representing about 50% of market share). The two groups have also implemented dedicated climate strategies since 2004. We outline the development and main characteristics of these strategies below.

Corporate climate strategies and public climate policy: two main phases

Phase 1 – voluntary agreement

Until 1990, the Swiss energy and CO_2 reduction policy was based on the voluntary program 'Energy 2000' and relied on private initiatives. Things began to change in the early 1990s with preparatory work for the UNFCCC, as the Swiss government expressed willingness to introduce a mandatory tax on CO_2 emissions (Lehman and Rieder 2002). After several national consultations and a political process during which corporate interest groups and right-wing political parties mostly opposed a mandatory tax, the government introduced a new CO_2 act: it would enact a tax only if voluntary initiatives failed to reach the Kyoto Protocol targets (−8% during 2008–2012 with respect to baseline year 1990). Furthermore, the government planned to exempt energy intensive corporations from a future tax, if they voluntarily agreed to CO_2 reduction objectives.

Following the entry into force of the CO_2 act in 2000, Coop and Migros progressively intensified their climate protection efforts. In the first half of 2004, they were among the very first companies to sign a voluntary agreement to reduce GHG emissions with the private agency 'EnAW' – an emanation of the peak business association 'EconomieSuisse' supporting those firms willing to commit to voluntary CO_2 reduction commitments.

Phase 2 – overcompliance with legally binding commitments

Meanwhile, it became clear to the Government that the voluntary programme would be insufficient to meet the Kyoto objectives. In 2004, it initiated steps to enact a mandatory CO_2 tax on fossil fuels (Ingold 2008).

The main political parties and interest groups were clearly divided on whether or not to support this tax. Coop and Migros, like some other big companies which fulfilled the conditions to be exonerated, voiced their support – in contrast to EconomieSuisse. A lengthy and conflictual political process ended in 2007 with the decision to introduce a CO_2 tax from 2008 onwards, but solely on heating fuels and not on motor fuels. At this point in time, firms that agreed to a legally binding reduction objectives were officially exempted. A financial sanction was foreseen in case of non-attainment of the objectives: the offending firm would have had to redeem the exonerated tax total for the years 2008–2012.

The debates around the CO_2 tax coincided with a boost in Coop and Migros corporate climate strategies. The two firms multiplied actions in relation to climate protection ranging from installing renewable energy sources to labelling climate-friendly retail products. The information campaign on their climate efforts was reinforced too. Both companies announced (in 2008 for Coop, in 2011 for Migros) that they were willing to voluntarily abate their CO_2 emissions well beyond the legal requirements.

In 2012, the CO_2 act was revised in order to account for the end of the first period of the Kyoto Protocol. Coop and Migros renegotiated and renewed their formal reduction engagements. At about the same time, the government created a working group on environmental labels and proposed a draft revision of the Federal Act on environmental protection that included an obligation to transparently inform consumers about the ecological and climate impacts of retail products. Coop and Migros opposed this new legislation, which the parliament discarded in 2014.

Applying the typology: the three faces of corporate climate strategies

Internal face

The reduction pledges voluntary agreed by Coop and Migros in 2004 were similar in ambition: 16% of GHG emissions reduction in 2010, compared to a business as usual scenario. Both companies strongly reinforced these objectives after the passing of the CO_2 tax. Coop set the goal of becoming climate-neutral in 2023, with a 50% cut in emissions compared to 2008 and compensation of the rest by acquiring Kyoto certificates. Migros, in turn, announced a somewhat less impressive abatement of 20% in 2020 (compared with 2010).

Both companies mainly focused on their direct emissions (Scope 1), and deployed a large spectrum of actions to achieve significant abatement (buildings insulation, phasing out of HCFCs, eco-lighting, renewable energy generation, carbon neutral transportation policy). The two companies also acted proactively on Scope 2 emissions, increasingly relying on hydropower. Scope 3 emissions seemed a more problematic challenge. The expert agency EnAW did not provide support on such emissions, and the CO_2 act did not oblige firms to monitor nor to control them. Coop and Migros only timidly engaged in carbon-labelling and assortment limitations to reduce the carbon footprint of products. Coop (2012, p. 24) justified this reluctance as due to the complexity of carbon labelling for *'average citizens'*, implementation costs, and the lack of reliability of methods for accounting indirect GHG emissions. Nevertheless, the firms introduced two labels:

the label 'By Air', indicating that the emissions related to air transportation had been compensated for; and the label 'Climatop', awarded to products with the lowest carbon footprint for a given product category. Among the different eco-labels advertised by Coop and Migros, the climate labels proved least commercially successful. Furthermore, the life-cycle assessment methodology used to deliver the Climatop label resulted in comparatively bad evaluation of local products in comparison to overseas organic products, one of the factors leading Migros to abandon the label in 2014.

Discursive face
Since 2004, Coop and Migros developed an intense, planned and coherent communication of their climate efforts. This consisted of numerous environmental reports, commercial advertisements, and marketing campaigns. Coop, for instance, was very proactive in publicising its 'spirit of initiative' and creativity in developing strategies that went beyond regulatory requirements (e.g., Coop 2005, p. 16), carefully listing and communicating climate measures. Coop has listed more than 300 environmental 'acts', while Migros has made more than 60 'promises'.

Interestingly, this marketing made no distinction with regards to the environmental impacts or the legally binding nature of the measures advertised. The goal was mainly to draw the public's attention to the exemplariness of the company compared to the competition. Coop and Migros also fully engaged in a tit-for-tat battle: practically every announcement of a green initiative by one was answered with a similar engagement by the other. Coop, however, dealt a fatal blow with the campaign around its CO_2 neutrality strategy in 2008, which Migros has yet to match in terms of environmental merits.

Political face
On the political front, the attitude of the two companies varied across time and sectors. The two firms supported a CO_2 levy that would give them a comparative advantage in the market and put a burden on their main contenders. In parallel, Coop and Migros positioned themselves as exemplary businesses that did not need any steering from the state to protect the climate. Coop was quite explicit: *'taking up ambitious commitments on a purely voluntary basis allows avoiding a multiplication of legislation'* (2005, p. 72, our translation) – although, as we have seen, these commitments were not taken on a purely voluntary basis.

From 2004 onwards, Coop and Migros constantly voiced support for the several revisions of the climate policy framework. The opposition of both firms to a strengthening of public regulation on environmental information and labelling of products constituted, however, a noteworthy exception.

Discussion

Evaluating climate leadership: Coop and Migros climate behaviour reassessed

When selecting our case studies, we assumed that Coop and Migros would exemplify climate leadership. The application of our typology allows us to precisely analyse this inference by disentangling the different 'faces' of their strategies.

Coop and Migros acted as *pushers* for most of the observed period. They adopted and translated ambitious climate protection objectives into numerous measures (internal face), established advanced and coherent marketing of climate efforts (discursive face), and lobbied in favour of environmental regulation (political face), except with regards to climate labelling of products.

On a closer look, however, we must contextualise the robustness of our typology. First, the strategic behaviour of the firm in relation to climate change is not unequivocal, but seems to vary depending on the issue at stake. For instance, the two companies acted as pushers with regards to Scope 1 and 2 emissions, but oscillated between the positions of free riders and pioneers respectively to Scope 3 emissions. When Coop and Migros introduced the climate labels 'by air' and 'climatop', they were, indeed, among the first companies to do so, along with Tesco in the United Kingdom and Casino in France. However, these climate labels were never implemented at full-scale or remained in the testing phase. If we could consider Coop and Migros as pioneers in a first phase, their later behaviour approximates that of a free rider, waiting for other European or global players to take the lead.

Secondly, the typology does not allow us to make absolute judgements regarding the merits of corporate climate strategy. The classification is only applicable relative to a given set of corporate actors. For instance, if we can consider both Coop and Migros as climate pushers relative to the whole population of food retailers, Coop is a greater pusher than Migros; the CO_2 objectives announced by Coop in 2008 (climate neutrality) were more ambitious than and preceded by three years those of Migros.

Ultimately, if the typology proved useful to classify and decipher corporate strategies, this kind of heuristic device suffers from an inherent limitation: it is context-dependent. First, with regards to the geographical scale: we can consider one a climate leader in a defined region, but a laggard compared to actors in other places or at other levels (world, region, city). Second, relative to *the sector*: some firms may be exemplary, whereas the sector as a whole could be lagging. Finally, the typology is *time dependent*: propensity to act as climate leader may vary according to various conjunctures and events, and the laggards of today may very well be the pioneers of tomorrow. Such external parameters need accounting for when applying the typology to ensure consistent evaluations and comparisons between the cases.

Theory building: the pathway to climate leadership

The analysis of the climate protection strategies of Coop and Migros reveals a common pathway to climate leadership and also certain divergences. Figure 4 illustrates the causal mechanism identified in both cases, based on the three building blocks our exploratory framework proposes.

Motives and capacities

When the State started regulating energy efficiency in the early 1990s, first with a voluntary approach, Coop and Migros were already ahead of the pack, having adopted internal energy targets during the late 1970s. Such pre-existing engagement in green initiatives can only be explained by the firms' internal motives and capacities.

In both the boards of Migros and Coop, key individuals developed a strong sense of concern for ecology and believed in the market potential of green initiatives. The governance structure of both firms allowed these key individuals to play the role of opinion leaders and to push for environmental exemplarity.

We can hardly separate these individual beliefs and values from a perception of the firm's self-interest, as green initiatives in general were expected to hold a market potential, to bring about reputation benefits and to allow for growing margins on product sales. This is particularly true in the case of Coop. Environmental exemplariness represented the key element of its development and market positioning since the 1990s.

Individual leadership, ecological values and converging interests would not suffice, however, to explain the market positioning of the firms, had their resources and capacities been lower. First, Coop and Migros disposed of significant financial and technical means due to their dominant position in the Swiss food market. Second, their decentralised structures, covering the whole country, allowed them to invest and experiment with green initiatives in local supermarkets, before upgrading and up-scaling them to the national level. In that respect, however, the more centralised structure of Coop seems to have been an important factor to explain its more ambitious climate strategy. The head of the company was able to implement the climate objectives and measures decided at the top, whereas the more decentralised governance structure of Migros complicated such top-down processes.

Our results suggest that availability of sufficient resources and the existence of environmental motivations are a necessary first step for a pathway towards climate leadership to develop. This implies, however, that corporate climate leadership could stay out of reach for small companies with only limited resources, and be restricted either to niche actors endowed with pioneer reduction technologies or to major firms with large resources (Biedenkopf et al. 2019 – this volume).

Social interactions

A very specific web of social interactions set into motion the pre-existing motives and capacities of the two firms. Until the early 2000s, Coop and Migros were in a duopolistic situation. They benefitted from strong regional structures and a wide base of relatively captivated consumers. This implied intense competition between them. Environmental protection was one of the fields where this took place. If both Coop and Migros saw important market potential in the growing fraction of environmental-friendly consumers, Coop decided to seize this niche in the hope of catching up with the higher market share of its main competitor.

The entry into force of the CO_2 act in 2000 shortly preceded the arrival of new competitors: the German hard discounters Lidl and Aldi. In this context of increased competition, Coop and Migros perceived climate protection as a way to further distinguish themselves not only from each other, but also from their direct challengers. Climate laws contributed to push for the development of novel environmental solutions, whereas the market situation pushed Coop and Migros to engage further in climate leadership. Being ahead in terms of environmental commitments bestowed on Coop and Migros a persuasive argument for why consumers should stay faithful despite their being generally more expensive than the German discounters. Furthermore, instead of reducing product prices, specialising in environmental-friendly products offered Coop and Migros a way to resist the new competitors while increasing the profit margin on sales.

Social interaction surrounding the adoption of corporate climate strategies also included the influence of third-party actors. Coop and Migros worked intensively with the parapublic agency EnAW. They also initiated economic collaborations with 'climate businesses', that is firms that provide products or services that help to calculate or to reduce emissions. Moreover, Coop and Migros established partnerships with the WWF. In exchange for the promise to reach certain environmental objectives and with the payment of a substantial amount of money, the WWF would positively communicate on the efforts accomplished. These forms of cooperation are quite representative of the move away from confrontation towards the non-conflictual relationships with business interests that some transnational NGOs such as WWF engaged in (Mol 2000). Altogether, energy or climate agencies, climate businesses and environmental organisations – working more or less closely with environmental public administration – constituted a new 'climate coalition', acting as a social force in business relationships. Through lobbying efforts, consulting or products and services delivery, this coalition contributed to push firms to adopt exemplary behaviour regarding climate protection.

Regulatory framework

Although both firms claim to develop purely voluntary climate strategies, we observe that the regulatory framework clearly contributed to shaping them. Most of the concrete measures announced by the two retailers can, indeed, either be related to the legally binding reduction commitment negotiated with the EnAW and monitored by the federal administration or to other legal obligations (e.g. the legal ban of CFC and HCFC). On the contrary, their actions proved to be far less ambitious in fields where no public requirements existed (e.g. on Scope 3 emissions).

The unique features of Swiss climate policy, based on both voluntarism (negotiation of legally-binding CO_2 reduction commitments) and threats (tax), strongly incentivised firms to take voluntary action first, in order to avoid the sanction of a tax and the risk of more stringent regulations, and, second, to avoid the reputational risk of appearing publicly as a laggard.

The importance of the CO_2 act becomes obvious when looking at the timing of corporate actions (Figure 3). The firms took the most significant measures in close anticipation or reaction to the introduction of the CO_2 act and its further revision. Actually, Coop and Migros began communicating and labelling their action as CO_2 reduction or climate protection initiatives only after the act's passage.

The CO_2 act did not only pressure firms to take the lead on climate protection; it also opened new market opportunities. Although largely provoked by a change in state regulations, one of the main objectives of the communication campaigns was to emphasise the voluntary and exemplary aspects of corporate actions. The firms turned the early efforts of Coop and Migros to go beyond legal expectations into a commercial advantage. Pioneering in CO_2 reduction initiatives allowed Coop and Migros to anticipate regulatory risks, but also to increase their competitiveness against new contenders. By doing so, the CO_2 act contributed to both the ecologisation of the Swiss food market and the strengthening of domestic corporate actors confronted by outsiders.

Whereas the Porter hypothesis emphasises the link between flexible regulation and corporate climate leadership, we have provided a more precise and nuanced account of the conditions under which state regulations might foster climate leadership. In the case of Swiss food retailers, we observe a relationship between flexible climate policies and the emergence of corporate climate leaders. Leaders need policy standards to over-comply with in order to publicly appear as being ahead of the pack. Furthermore, climate policy can incentivise firms to act as climate leader, provided it allows them to fully capture the reputational merits of climate protection measures. This occurs

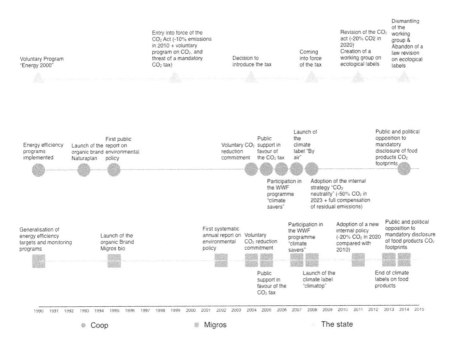

Figure 3. Timeline of principal climate-relevant decisions.

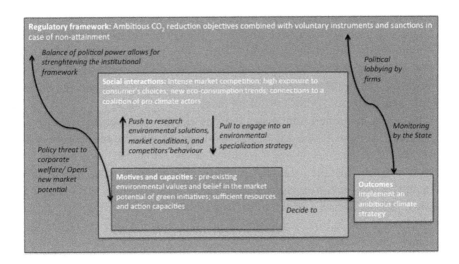

Figure 4. The mechanism of corporate climate leadership.

when firms are in a position to advertise any climate protection measures as resulting from their own private and voluntary initiative, even in cases where the firms took measures to comply with policy-defined objectives.

We highlight a mix of three combined features of climate policy that facilitate corporate climate leadership. We argue, based on our observations, that the policy design best suited to promote corporate climate leadership would rely on *ambitious and stringent policy objectives* setting sector-specific standards, associated with *flexible instruments* giving leeway to private and voluntary initiatives, backed up by *public monitoring and sanctions* in case of goal non-attainment. Such a policy mix creates a social space in which 'the shadow of the hierarchy' (Scharpf 1997) crucially contributes to make corporate actors responsible for climate protection measures, rewarding leaders and penalising laggards.

Conclusion: corporate actors' role in polycentric climate governance

We introduced a typology to analyse corporate behaviours towards climate change and proposed a framework to identify the mechanisms behind these strategic choices. We applied these conceptual tools to two Swiss food retailers, which we expected to be typical of climate leaders evolving in a context of flexible regulations.

We can designate both firms as pushers in the current polycentric climate governance system, notwithstanding the carbon footprint of food products, a domain where they tended to act in-between pioneers and free riders. We have demonstrated that our typology of corporate behaviours is a useful heuristic device to qualify firms' strategies on climate change. A systematic classification of private actions on climate remains complex, however, due to the multifaceted aspects of firms' strategies and the inherently static nature of this kind of typology.

We also identified the main features of the mechanism relating to corporate climate leadership. Some components of this mechanism may be specific to the Swiss context, for instance the duopolistic structure of the food market and the strength of eco-consumption trends, which have enabled a race towards climate exemplarity. Nevertheless, we argue that other elements clarify the Porter hypothesis and the conditions under which corporate actors could come to play the role of leaders in polycentric climate governance.

Corporate leaders need a regulatory framework that forms the baseline against which they can demonstrate their exemplariness, and in relation to which they can obtain a comparative advantage. In the two cases in point, the state established this structure of incentives through a complex climate policy that featured stringent goals associated with flexible instruments, and sanctions in case of non-attainment. If these findings seem to emphasise the need to reconsider the role of the national state to enable private climate innovation, this does not need to be the case. In the current polycentric climate governance, ambitious goals, flexible means and sanctions could theoretically be the features of institutional arrangements designed at the global, the national, the regional, the local levels or even between private actors. The question is not anymore where and by whom, but rather how and when.

Note

1. IFOAM EU Group. 2016. Organic in Europe: Prospects and development 2016. Brussels. https://shop.fibl.org/chen/mwdownloads/download/link/id/767/

Acknowledgements

This work was supported by Innovations in Climate Governance (INOGOV), the Swiss National Science Foundation (SNSF) in the frame of the project *The Governance of Climate change Adaptation*, grant number 100017_153525, and the national research programs *Healthy nutrition and sustainable food production* (NRP 69), grant number 406940_145181.

Disclosure statement

No potential conflict of interest was reported by the authors.

Funding

This work was supported by Innovations in Climate Governance (INOGOV), the Swiss National Science Foundation (SNSF) in the frame of the project The Governance of Climate change Adaptation, grant number 100017_153525, and the national research programs Healthy nutrition and sustainable food production (NRP 69), grant number 406940_145181.

ORCID

Johann Dupuis http://orcid.org/0000-0001-5098-7956

References

Altman, M., 2005. The ethical economy and competitive markets: reconciling altruistic, moralistic, and ethical behavior with the rational economic agent and competitive markets. *Journal of Economic Psychology*, 26 (5), 732–757. doi:10.1016/j.joep.2005.06.004.

Anckar, C., 2008. On the applicability of the most similar systems design and the most different systems design in comparative research. *International Journal of Social Research Methodology*, 11 (5), 389–401. doi:10.1080/13645570701401552.

Andonova, L.B., Betsill, M.M., and Bulkeley, H., 2009. Transnational climate governance. *Global Environmental Politics*, 9 (2), 52–73. doi:10.1162/glep.2009.9.2.52.

Baranzini, A., Thalmann, P., and Gonseth, C., 2004. Swiss climate policy: Combining VAs with other instruments under the menace of a tax. *Voluntary approaches in climate policy*. Cheltenham, UK, Northampton, MA, USA: E. Elgar. 249–276.

Biedenkopf, K., Eynde, S.V., and Bachus, K., 2019. Environmental, climate and social leadership of small enterprises: fairphone's step-by-step approach. *Environmental Politics*, 28, 1.

Burch, D. and Lawrence, G., 2005. Supermarket own brands, supply chains and the transformation of the agri-food system. *International Journal of Sociology of Agriculture and Food*, 13 (1), 1–18.

Coop, 2005. *Durabilité. Faits et chiffres relatifs à l'évolution économique, écologique et sociale du groupe Coop*. Basel: Coop Genossenschaft.

Coop, 2012. *Rapport sur le développement durable 2011*. Basel: Coop Genossenschaft.

Darnall, N., 2009. Regulatory stringency, green production offsets, and organizations' financial performance. *Public Administration Review*, 69 (3), 418–434. doi:10.1111/puar.2009.69.issue-3.

Downie, J. and Stubbs, W., 2013. Evaluation of Australian companies' scope 3 greenhouse gas emissions assessments. *Journal of Cleaner Production*, 56, 156–163. doi:10.1016/j.jclepro.2011.09.010.

Dupuis, J. and Knoepfel, P., 2015. *The politics of contaminated sites management*. Cham: Springer.

Fligstein, Neil, 2001. Social skill and the theory of fields. *Sociological Theory*, 19 (2), 105-125. doi: 10.1111/0735-2751.00132

Hale, T., 2016. "All hands on deck": the Paris agreement and nonstate climate action. *Global Environmental Politics*, 16 (3), 12–22. doi:10.1162/GLEP_a_00362.

Hill, M. and Hupe, P., 2014. *Implementing public policy: an introduction to the study of operational governance*. London: Sage.

Ingold, K., ed., 2008. *Analyse des mécanismes de décision: le cas de la politique climatique suisse*. Zürich: Rueggger Verl.

Jaffe, A.B. and Palmer, K., 1997. Environmental regulation and innovation: a panel data study. *The Review of Economics and Statistics*, 79 (4), 610–619. doi:10.1162/003465397557196.

Jordan, et al., 2015. Emergence of polycentric climate governance and its future prospects. *Nature Climate Change*, 5 (11), 977–982.

Knoepfel, P., et al., 2011. *Public policy analysis*. Bristol: The Policy Press.

Koehler, D.A., 2007. The effectiveness of voluntary environmental programs. *Policy Studies Journal*, 35 (4), 689–722. doi:10.1111/j.1541-0072.2007.00244.x.

Korpi, W., 1985. Power resources approach vs. action and conflict: on causal and intentional explanations in the study of power. *Sociological Theory*, 3 (2), 31–45. doi:10.2307/202223.

Koski, C. and May, P.J., 2006. Interests and implementation: fostering voluntary regulatory actions. *Journal of Public Administration Research and Theory*, 16 (3), 329–349. doi:10.1093/jopart/mui048.

Lanoie, P., *et al.* 2011. Environmental policy, innovation and performance: new insights on the Porter hypothesis. *Journal of Economics & Management Strategy*, 20 (3), 803–842. doi:10.1111/j.1530-9134.2011.00301.x.

Lantos, G.P., 2002. The ethicality of altruistic corporate social responsibility. *Journal of Consumer Marketing*, 19 (3), 205–232. doi:10.1108/07363760210426049.

Lehman, L. and Rieder, S., 2002. *Wissenschaftliches wissen in der politischen auseinandersetzung: fallstudie zur genese des CO_2 gesetzes im auftrag der arbeitsgruppe transdiziplinarität der energiekommission der Schweizerischen akademie der technischen wissenschaft (SATW)*. Luzern: Interface.

Liefferink, D. and Wurzel, R.K.W., 2017. Environmental leaders and pioneers: agents of change? *Journal of European Public Policy*, 24 (7), 651–668. doi:10.1080/13501763.2016.1161657.

Lubin, D.A. and Esty, D.C., 2010. The sustainability imperative. *Harvard Business Review*, 88 (5), 42–50.

Lyon, T.P. and Maxwell, J.W., 2008. Corporate social responsibility and the environment: a theoretical perspective. *Review of Environmental Economics and Policy*, 2 (2), 240–260. doi:10.1093/reep/ren004.

Lyon, T.P. and Montgomery, A.W., 2015. The means and end of greenwash. *Organization & Environment*, 28 (2), 223–249. doi:10.1177/1086026615575332.

Matsumura, E.M., Prakash, R., and Vera-Muñoz, S.C., 2014. Firm-value effects of carbon emissions and carbon disclosures. *The Accounting Review*, 89 (2), 695–724. doi:10.2308/accr-50629.

Maxwell, J. and Briscoe, F., 1997. There's money in the air: the CFC ban and DuPont's regulatory strategy. *Business Strategy and the Environment*, 6 (5), 276–286. doi:10.1002/(ISSN)1099-0836.

McMichael, P., 2009. A food regime genealogy. *The Journal of Peasant Studies*, 36 (1), 139–169. doi:10.1080/03066150902820354.

Mol, A.P.J., 2000. The environmental movement in an era of ecological modernisation. *Geoforum*, 31 (1), 45–56. doi:10.1016/S0016-7185(99)00043-3.

Nesta, L., Vona, F., and Nicolli, F., 2014. Environmental policies, competition and innovation in renewable energy. *Journal of Environmental Economics and Management*, 67 (3), 396–411. doi:10.1016/j.jeem.2014.01.001.

Nordhaus, W., 2015. Climate clubs: overcoming free-riding in international climate policy. *American Economic Review*, 105 (4), 1339–1370. doi:10.1257/aer.15000001.

Oekom Research AG, 2018. *Swiss retailers leading the way in sustainability management* [online]. Available from: http://www.oekom-research.com/index_en.php?content=pressemitteilung_20062011 [Accessed 1 May 2018].

OFEV, 2015. *Comment réduire les émissions de CO_2: dix exemples d'innovations* [online]. Available from: https://www.bafu.admin.ch/bafu/fr/home/themes/climat/dossiers/conference-paris-cop21-climat/comment-reduire-les-emissions-de-co2-dix-exemples-dinnovations.html--611154117 [Accessed 20 May 2017].

Olivier, J.G.J., *et al.*, 2014. *Trends in global CO_2 emissions: 2013 report*. PBL Netherlands Environmental Assessment Agency and EC JRC Institute for Environment and Sustainability. The Hauge.

Palmer, K., Oates, W.E., and Portney, P.R., 1995. Tightening environmental standards: the benefit-cost or the no-cost paradigm? *Journal of Economic Perspectives*, 9 (4), 119–132. doi:10.1257/jep.9.4.119.

Pizer, W.A., Morgenstern, R., and Shih, J.-S., 2011. The performance of industrial sector voluntary climate programs: climate wise and 1605(b). *Energy Policy*, 39 (12), 7907–7916. doi:10.1016/j.enpol.2011.09.040.

Porter, M.E. and Van der Linde, C., 1995. Toward a new conception of the environment-competitiveness relationship. *The Journal of Economic Perspectives*, 9 (4), 97–118. doi:10.1257/jep.9.4.97.

Poteete, A.R., Janssen, M.A., and Ostrom, E., 2010. *Working together: collective action, the commons, and multiple methods in practice*. Princeton: Princeton University Press.

Rhodes, R.A.W., 2007. Understanding governance: ten years on. *Organization Studies*, 28 (8), 1243–1264. doi:10.1177/0170840607076586.

Scott, W. Richard, and John W Meyer, eds. 1994. *Institutional environments and organizations: Structural complexity and individualism*. Thousand Oaks: Sage.

Sabatier, P.A. and Weible, C.M., 2014. *Theories of the policy process*. 3rd ed. Boulder, CO.: Westview.

Scharpf, F., 1997. *Games real actors play: actor-centered institutionalism in policy research*. Oxford: Westview.

Schweizer, R., 2015. Law activation strategies (LAS) in environmental policymaking: a social mechanism for re-politicization? *European Policy Analysis*, 1 (2), 132–154. doi:10.18278/epa.1.2.7.

Seawright, J. and Gerring, J., 2008. Case selection techniques in case study research. *Political Research Quarterly*, 61 (2), 294–308. doi:10.1177/1065912907313077.

Stroup, S.S. and Wong, W.H., 2018. Authority, strategy, and influence: environmental INGOs in comparative perspective. *Environmental Politics*, [online] 27 (6).

Vandenbergh, M.P., 2007. The new Wal-Mart effect: the role of private contracting in global governance. *UCLA Law Review*, 54 (4), 913–970.

Vormedal, I., 2011. From foe to friend? Business, the tipping point and US climate politics. *Business and Politics*, 13 (3), 1–29. doi:10.2202/1469-3569.1350.

Wright, C. and Nyberg, D., 2014. Creative self-destruction: corporate responses to climate change as political myths. *Environmental Politics*, 23 (2), 205–223. doi:10.1080/09644016.2013.867175.

Wurzel, R., Liefferink, D., and Torney, D., 2019. Pioneers, leaders and followers in multilevel and polycentric climate governance. *Environmental Politics*, 28, 1.

Yalabik, B. and Fairchild, R.J., 2011. Customer, regulatory, and competitive pressure as drivers of environmental innovation. *International Journal of Production Economics*, 131 (2), 519–527. doi:10.1016/j.ijpe.2011.01.020.

ə OPEN ACCESS

The oil and gas sector: from climate laggard to climate leader?

Matthew Bach

ABSTRACT
The oil and gas industry has traditionally been reticent to engage with the issues surrounding climate change, typically being cast as a laggard. Yet, over recent years, the sector has begun taking on a more active role in climate governance, doing so in a variety of capacities – as initiators, catalysts and participants in industry-led or multi-stakeholder efforts. The Oil and Gas Climate Initiative is reviewed, as a case study to illustrate emerging climate leadership within the global oil and gas industry. In 2015, its members committed to a two-degree pathway. The paucity of research on the nascent role of oil and gas firms in climate governance is addressed.

Introduction

'Fossils', 'merchants of doubt', 'polluters' are the terms that we mostly associate with the oil and gas industry in relation to climate change. Indeed, the industry has done much to deserve such labels. As early as 1989, just 1 year after the establishment of the first global climate-dedicated organization, the Intergovernmental Panel on Climate Change (IPCC), a number of the largest oil and gas and coal firms came together to launch the Global Climate Coalition (GCC), an advocacy group dedicated to promoting climate change scepticism (Brown 2000, Revkin 2009).

Yet, within a decade of the GCC's founding, cracks in the oil and gas sector's stance towards climate change became apparent: certain companies, including Shell and BP, two of the largest EU-based international oil companies (IOCs), broke with the organization and aligned themselves with the climate science of the IPCC; others, such as ExxonMobil and Chevron, maintained a hard line against climate science and policy (Levy and Kolk 2002). The lengths to which some firms went to discredit climate science have been exposed, as in the case of ExxonMobil (Jennings et al. 2015; Jerving et al. 2015, Liebermann and Rust 2015, Banerjee et al. 2015a, 2015b).

This is an Open Access article distributed under the terms of the Creative Commons Attribution-NonCommercial-NoDerivatives License (http://creativecommons.org/licenses/by-nc-nd/4.0/), which permits non-commercial re-use, distribution, and reproduction in any medium, provided the original work is properly cited, and is not altered, transformed, or built upon in any way.

Even if some firms were willing to acknowledge climate science and policy, a more general reluctance to become involved in global efforts to address climate change persisted until recently. An interviewee from a state-owned oil and gas firm argues that the 'arm's length' approach has begun to give way to more active industry involvement, one in which climate change is no longer taken as a pure negative, but is instead seen as an opportunity for the creation of new business units, the development of new technologies and the optimization of existing processes. The majority of interviewees from industry echoed this view; one went so far as to say that they intend to be 'carbon competitive'. Eikeland and Skjærseth (2019 – this volume), in their study of oil and power industries' responses to EU emissions trading, further corroborate this sense in noting certain companies' 'proactive response strategies'.

This shift can be traced to wider changes, from the rapid rollout of renewable energy technologies to the diffusion of new concepts such as stranded assets (Griffin et al. 2015, van der Ploeg 2016). Of these, the growing involvement of non-state actors in governance processes is particularly significant (Downie 2014, Bäckstrand et al. 2017). As regards firms, this represents a momentous shift: the commonly held view that sustainability represents a challenge to economic growth (Braithwaite and Drahos 2000), and by extension to conventional business models, has become a thing of the past. Major companies have committed to ambitious emissions reduction strategies, and others have banded together in support of climate change mitigation and adaptation (Chan et al. 2015, UNEP 2015, Pattberg and Widerberg 2016).

For the oil and gas industry, the emerging post-Paris world speaks against avoidance as a strategy, especially as the potential cost of inaction swells. Indeed, whereas the world previously lacked credible alternatives to fossil fuels, a future no longer dependent on them has come into sight (e.g. Holland et al. 2016, Papaefthymiou and Dragoon 2016). Firms are now having to consider the very real possibility that oil and gas will no longer play a central role in our energy systems and, worse yet, that they could become locked out of a low-carbon future. Moreover, these changes are threatening the very business models that underpin many of the largest companies (Mitchell et al. 2015, Stevens 2016).

It is in this context that a number of climate-focused governance initiatives have emerged within the sector (Nasiritousi 2017). The earliest, the Oil and Gas Climate Initiatives (OGCI) is a club that brings together 10 of the largest firms, who account for close to 20 per cent of global oil and gas production, as well as nearly 12 per cent of historical greenhouse gas emissions (GHGE). Founded in 2014, and in many ways just starting up, its members have voiced their support for current climate policy and science and have begun to reimagine themselves within a low-carbon energy future.

Though oil and gas firms have become involved in other initiatives[1], here I focus on the OGCI for three reasons: first, it is the only group entirely constituted by industry; second, its members have significant structural power (and therefore, as we will see below, structural leadership potential) – financially, technologically, and also through their GHGE; and third, it has a varied membership, spanning both national and international oil companies in developed and developing contexts, which has the potential to provide clues regarding the convergence or divergence of positions taken by firms within this industry.

Here, I propose to explore: to what extent and under what conditions do oil and gas companies act as climate leaders or pioneers? In searching out answers, I make use of the conceptual framework proposed by Liefferink and Wurzel (2017) to analyse the roles of actors in environmental governance in terms of their positions, motivations, styles and strategies.

The second section offers an overview of the methodological issues encountered in the study of a nascent initiative in the oil and gas sector, as well as some of the conceptual ones related to the use of the framework outlined above. I then present the OGCI in a third section, followed by a discussion of the extent to which, and the conditions under which, the initiative may be acting as a climate leader or pioneer. In conclusion, I reflect on the importance of examining the OGCI, as well as on some of the conceptual limitations of the Liefferink and Wurzel (2017) framework, before attempting to tease out some of this contribution's broader implications for climate governance.

Methods

Examining the extent to which oil and gas companies act as leaders and the conditions under which this can take place poses several methodological challenges. The first relates to the nature of the selected case, the OGCI, which has only been founded relatively recently, making it challenging to trace processes and provide a longitudinal vision. The framework that Liefferink and Wurzel (2017) put forward assists in this regard by offering a typology built around the concepts of leader and laggard.

The second challenge directly links to the conceptual framework itself, which seeks to distinguish between the internal and external faces of an actor's environmental ambitions, as a means of determining their position within the leader-laggard typology. This typology was initially put forward in the context of nation-states, and so it is less clear what the specific criteria might be when applied to non-state actors such as firms. This is especially the case in relation to the internal dimension. The majority of publicly available documents and even the statements of interviewees primarily aim at an external audience and may not necessarily genuinely represent internal ambitions.

Furthermore, the OGCI as a nascent initiative has yet to produce extensive materials for analysis, which makes the collection of empirical data all the more crucial. This has been dealt with by combining the analysis of varied sources – websites, videos, media articles, trade publications, social media feeds – with semi-structured interviews carried out between January and June 2016. These involved senior managers from the OGCI and from the firms taking part in it. In order to widen the view somewhat, I carried out additional interviews from January to July 2017 with external stakeholders, whether industry associations or civil society groups. A related difficulty is specifically due to the nature of the oil and gas sector. Given the traditional reluctance of the sector to engage in anything related to climate, we may simply not have the tools to say to what extent these efforts are made in good faith and if they will bear fruit. For this reason, I focus more on the conditions under which the positions of actors occur, rather than speculate about underlying motivations.

The case: the OGCI

In January 2014, the chief executives from some of the world's largest oil and gas firms gathered on the sidelines of the World Economic Forum in Davos, Switzerland. They were increasingly concerned about the risks that climate change poses and wanted to position themselves as 'recognized and ambitious provider[s] of practical solutions for climate change mitigation' (OGCI 2015). In short, they wanted to be on the right side of a carbon-constrained future.

Led by the Chairman of Saudi Aramco and the CEOs of Total and Eni, the conversation grew into a set of commitments from major national and international oil companies[2], most notably to support a two-degree target. In September 2014, Khalid A. Al-Falih, the President and CEO of Saudi Aramco, officially announced the OGCI at a UN Climate Summit in New York, portraying it as 'an example of how the oil and gas industry is positioning itself once again to be the key provider of solutions to global energy challenges' (UN 2014).

A semi-public technical workshop in Paris in May 2015 first lifted the veil of the inner workings of the OGCI. Led by Gerard Moutet, the chair of the OGCI executive committee and the vice-president of Total for climate and energy, the event gathered views from member companies and selected stakeholders in support of the initiative's development. It also solicited feedback on a draft report detailing its agenda. In addition, this meeting initiated three 'work streams': long-term solutions, the role of natural gas, and instruments and tools for carbon reduction.

One of the goals was clearly to link the OGCI to more established actors in climate governance, notably by inviting two of the UN's highest authorities – Christiana Figueres (Executive Secretary of the UNFCCC) and Janos

Pasztor (Assistant UN Secretary General on Climate Change). OGCI also made a connection to the OECD's International Energy Agency (IEA) through its head, Fatih Birol.

It was only in October 2015 that the OGCI truly made it into the public eye, when the CEOs of its member companies gathered in Paris ahead of the UNFCCC's 21st Conference of the Parties (COP) to present the initiative and a first report setting out its ambitions (OGCI 2015). With such headlines as 'Oil bosses to meet in latest climate change offensive' (Bousso and Schaps 2015), this was an occasion for strong and optimistic statements on behalf of some of the world's largest GHG emitters (Politiques énergétiques 2015b).

Since then, the CEOs have continued to feature the initiative (e.g. Dudley 2015) and the OGCI published a press release welcoming the Paris Agreement. Although it might seem that the OGCI was primarily developed in anticipation of COP21, interviewees linked to the initiative stated their intention for it to last throughout the transition to a low-carbon future (20–30 years). Indeed, efforts to develop it continue apace, and those involved are optimistic that it will grow.

Much of the OGCI work in 2016 focused on internal coordination between the member companies. The major announcement of the year, however, came ahead of the UNFCCC COP in Marrakech (COP22) and in the form of a commitment from the OGCI members to invest USD 1 billion in low-emissions technologies. The majority of this fund would be used to accelerate the deployment of carbon capture, use and storage (CCUS) in gas-fired power plants, while also tackling the issue of methane leakages, which contribute significantly to climate change (Solsvik and Fouche 2016). A spokesperson for the group noted that the initiative would consider increasing their investment upon review (Sampathkumar 2016).

Positions, conditions, styles and strategies

Here, I probe the conditions under which the OGCI and its member oil and gas firms engage in leader-like behaviour, using the climate leaders and pioneers framework proposed by Liefferink and Wurzel (2017). More specifically, I do so by considering four key elements: positions, styles and strategies – derived from this framework – as well as conditions. I introduce this last element due to the difficulty of asserting the good faith of oil and gas actors in taking on leadership roles in relation to climate change and its governance. Indeed, in this context, considering the conditions under which these actors' positions occur may well be more fruitful than speculating about their underlying motivations.

Positions: internal and external

The first dimension of the framework that Liefferink and Wurzel (2017) put forward concerns the internal (environmental) ambitions of actors. In the OGCI's case, this means looking closely at the initiative's members, the 10 firms that have coordinated a common position on climate change.

As the initiative is CEO-led, there is a clear intention to provide internal leadership within the member companies. Members have highlighted this aspect as particularly important and the source of the initiative's power (Politiques énergétiques 2015b). This high-level involvement is, however, not purely symbolic, as the CEOs are not allowed to delegate their participation and must play an active part in the governance of the initiative, as well as in the choice of the topics pursued.[3] In an interview, a key manager from the initiative also noted that the CEO-centric design enables faster project planning and implementation for joint low-carbon efforts.

As detailed by senior managers from three member firms, the OGCI members are also investing resources beyond the leadership of their chief executives. Senior managers, mostly at the Vice-President level, form an executive committee, which is responsible for the day-to-day running of the initiative. Until March 2016, it held weekly teleconferences (now bi-weekly) and they continue to meet once every few months for at least a day. Company experts are also involved through working groups, which convene on a regular basis. This degree of activity clearly indicates internal commitment on the part of the member companies, as these would otherwise not be willing to expend such significant resources in terms of personnel and time.

Moreover, the 2016 decision to invest USD 100 million per firm further cements internal commitments. It could therefore be possible to portray the OGCI member companies as harbouring high internal environmental ambitions in relation to the OGCI, even if it remains too early to say whether these ambitions will crystalize into new and low-carbon practices, technologies and policies.

The second dimension that Liefferink and Wurzel (2017) offer to determine the position of actors is that of external (environmental) ambitions. We can take the governance interactions between the OGCI and other actors within the climate governance regime as a gauge for this. In particular, this means considering links with trade associations, states, international organizations and civil society.

The OGCI has complex ties with oil and gas trade associations: it makes use of guidelines that they have developed, for instance on climate reporting (cf. IPIECA, API and IOGP 2011), and nearly all individual firms are long-standing members.[4] The CEO of Total explicitly stated that 'it is better for climate change that we can advocate inside the [API] being a member, being around the table, we don't change our minds, my position is clear' (Politiques énergétiques 2015b). The CEO, speaking purely on behalf of Total, also declared: 'I'm committing the full company – all lobbyists from Total in all organizations have the same position' (Politiques énergétiques 2015b.). This signals a clear intention to provide external leadership at the level of the oil and gas sector via these trade organizations.

Interactions with states play a prominent role in the OGCI and unfold in two main ways. On the one hand, the initiative includes NOCs, which are part and parcel of states. There is thus a direct link between the OGCI and China, Saudi Arabia, Mexico and Norway. While some of these polities have been reluctant to engage in climate governance, the OGCI might offer a less visible arena for them to take on a more active role. On the other hand, the initiative explicitly seeks collaboration with states, for instance, by inviting former French Foreign Minister and chair of the COP21 negotiations Laurent Fabius to its October 2015 event. This desire to engage with states on climate governance indicates some degree of external ambition, yet it has not taken a prominent place in its actions.

When it comes to international organizations, there is a much clearer intention on the part of the OGCI to establish links, including with the UN, the OECD's IEA and the World Bank. Christiana Figuerres, the Executive Secretary of the UNFCCC was recorded as stating that the 'OGCI has a very important role to play in as much as OGCI has already accepted that we have to stay under 2C' (Politiques énergétiques 2015a). However, she went on to caution that 'we have to figure out how [to stay under 2C], while producing more energy, cheaper energy, cleaner energy, *and above all leaving fossil fuels underground*' (Ibid.; italics added). Thus, while the support of these actors seems to be extended in general terms only, there is an explicit attempt to provide some degree of external leadership vis-à-vis these fora.

The most challenging interactions to trace have been those with civil society. Two short video interviews with NGO representatives on the sidelines of the October 2015 event provide some empirical evidence of such interactions. Moreover, one interviewee highlighted that interactions are occurring more frequently than might be otherwise expected, though mostly via informal exchanges (Interview, 2016). Work is also being done with research organizations such as the Massachusetts Institute of Technology. Overall, however, it seems that the OGCI does not have significant external ambitions towards civil society.

Determining the position of the OGCI

Based on this review of the interactions within the OGCI and between it and other actors, including trade organizations, states, international organizations and civil society, it becomes possible to characterize the position of the initiative in terms of the Liefferink and Wurzel (2017) typology. The OGCI has strong internal environmental ambitions, expressed through the willingness of its members to commit resources and senior personnel; externally, it displays similar ambitions towards trade organizations and international organizations, though weaker ones towards states and low ones towards civil society. Table 1 (below) provides an overview of these ambitions and positions, and it highlights that one actor does not necessarily take a single position, which may vary contextually.

On the whole, if we are to make use of the Liefferink and Wurzel categories developed, this would imply that we could characterize the OGCI as a pusher, which is defined as an actor 'which takes the lead domestically and actively seeks to push other[s] to implement its internal ambitions' (Ibid., p. 4).

Conditions

The nascent involvement of the oil and gas sector in climate governance raises the important question of what conditions lead firms, which have either opposed or remained aloof from efforts to address climate change, to adopt more entrepreneurial positions. The question is, by extension, why are firms whose bread and butter is carbon-intensive suddenly claiming to take part in a global energy transition? Though this shift indeed warrants a healthy dose of scepticism, we cannot shrug it off as yet another instance of greenwashing. In fact, we can trace it to a number of macro-level changes occurring both within and outside of the climate regime.

For one, states are no longer the only governors of climate change; their prominence has declined as others (e.g. international organizations, civil society) have taken up calls for climate action (Pattberg 2016, see also Wurzel *et al.* 2019, – this volume). Nor have firms been absent from this bottom-up movement. Some have committed to ambitious emissions reduction strategies, and others have banded together in support of climate change mitigation and adaptation (Chan *et al.* 2015, UNEP 2015, Pattberg and Widerberg 2016).

We can link a second change to the fact that the blanks in the UN Framework Convention on Climate Change (UNFCCC) have begun to be

Table 1. Overview of ambitions and positions (adapted from Liefferink and Wurzel 2017).

	Internal ambitions	External ambitions			
		Trade assoc.	States	IOs	Civil society
High	x	x		x	
Neutral			X		
Low					x

filled through the COP process. States are taking on growing levels of commitment, for instance by agreeing to a two-degree Celsius pathway in Paris and going so far as to suggest a 1.5 degrees target. This has effectively created a boundary for societal actors, indicating the point beyond which they could expose themselves to sanctions. A central figure in the OGCI corroborated this view, citing a clearer vision of climate impacts and policies as an underlying motivation for the initiative.

Another development has been increasing support for a low-carbon future, in which energy is primarily generated from renewable sources such as wind, solar and hydro. The mainstreaming of renewable energy technologies as part of what has been called an energy transition makes it impossible to dismiss a low-carbon future as a pipedream: it is today a distinct possibility (Cherp *et al.* 2011, Verbong and Loorbach 2012, DDPP 2015). Oil and gas firms are following these changes closely and some interviewees from industry were quick to highlight how they themselves are keen to be a part of this transition. As one put it, 'we think we can play in [the renewable] segment' (Interview with a senior executive from a European oil and gas firm, 2016)

The scale of this transition has already translated into the overhaul of corporate strategies. In 2016, E.ON, one of Germany's largest energy firms, finished splitting from its fossil assets (Timperley 2016). That same year, the largest French utility Engie announced that it would aim to derive 90% of its earnings from low-carbon sources within three years (Froley 2016).

At the same time, discursive shifts have been occurring through which climate change is no longer thought of as a far-off threat, but plays a central part in current decision-making. Concepts such as the carbon bubble and stranded assets have pushed some large institutional investors to divest from their holdings in fossil energy (Mathieu 2015). Symbolically, the recent decision by the Rockefeller Foundation – descended of the very same Rockefeller who founded Standard Oil, the ancestor of major oil and gas firms such as Exxon and Chevron – to withdraw its investments from fossil companies made waves (RFF 2016). Nevertheless, industry interviewees questioned the importance of the divestment movement, suggesting that it is 'fundamentally flawed'.

Styles and strategies

In understanding the strategies through which the OGCI is engaging in climate governance, I have drawn on the Liefferink and Wurzel (2017) categories. Building on Young (2011), the strategies correspond to what they refer to as *types* of leadership and include the following: *structural*, which relates to the actor's material strength and capacity to affect systemic outcomes; *entrepreneurial*, which involves the negotiation of favourable results; *cognitive*, which refers to the actor's efforts to redefine positions through ideas, for instance, via scientific expertise; and *exemplary*, or leading by example. Analysing strategies through types of leadership has the advantage of highlighting what type of resources the actors mobilize in the pursuit of their goals. In this contribution, I have excluded exemplary leadership, as an example has yet to be set.

As for *styles*, their analysis permits not only a more detailed assessment of how these companies are approaching climate governance, but also to consider the intensity of their engagement and an indication of what kinds of impacts might be achieved. Liefferink and Wurzel (2017) point to the following styles: humdrum or *transactional*, relying on step-wise optimization which targets mostly short-term change; heroic or *transformational*, stemming from far-reaching and long-term ambitions which seek to bring about radical or even revolutionary change. Here, I will refer only to transactional or transformational styles as analytical concepts (see also Wurzel et al. 2019 – this volume).

Types of leadership

Based on the examination of the internal and external interactions of the OGCI, we can argue that it acts as a 'pusher', which 'takes the lead domestically and actively seeks to push [others] to follow its example' (Liefferink and Wurzel 2017, p. 954–5). Pushers act through two possible assemblages of leadership types (see Table 2 below).

The first option concerns actors capable of mustering structural leadership, and is highly relevant to the OGCI, as its members control vast resources and are responsible for a significant portion of historical GHGE. As an interviewee highlighted, oil and gas firms have a number of advantages in taking an active role in the energy transition: financial resources, know-how, and a significant innovation capacity (Interview, 2017). Nevertheless, in comparison to states or groups of states, we could see these firms as relatively minor players.

The second variety of pusher is less able to draw on structural leadership, but can instead combine its entrepreneurial and cognitive abilities. In this regard, the OGCI is clearly acting entrepreneurially by establishing the first

Table 2. Leadership types in different positions (adapted from Liefferink and Wurzel 2017).

Position	Type of leadership			
	Structural	*Entrepreneurial*	*Cognitive*	*Exemplary*
Laggard	-	-	-	-
Pioneer	-	-	-	X
Pusher (A)	x	(x)	(x)	(x)
Pusher (B)	(x)	x	X	(x)
Symbolic leader (A)	x	(x)	(x)	(x)
Symbolic leader (B)	(x)	x	X	-

climate governance initiative by and for industry; cognitively, it is putting forward a vision of a low-carbon oil and gas sector, which has the potential to influence future policy.

The OGCI therefore seems to be situated in between these two forms of pushers, as it is possesses qualities of structural leadership and also the abilities linked to entrepreneurial and cognitive elements. Though the OGCI uses the latter two types in all settings, we need to draw a distinction concerning the former (see Table 3). Indeed, within the oil and gas industry, it possesses a higher degree of structural leadership than it does towards states or groups of states. Overall, then, it would seem that the OGCI is strategically able to employ three types of leadership. What this implies, however, is less clear and will require further research.

Styles of leadership

Determining the styles of leadership as employed by the OGCI remains a fairly speculative affair for the moment since leadership styles – and especially those of a transformational nature – become defined only over time. Indeed, it is not yet possible to say whether the OGCI's leadership could take on a more radical dimension (i.e. a transformational style). However, as the initiative frames itself within a longer-term energy transition, we can exclude the transactional style that focuses on short-term actions. As such, it would appear that for the moment, the OGCI is making use of a transactional leadership style, which is primarily concerned with optimizing current systems to achieve its goals. Interviews with members of the OGCI so far seem to confirm this, for OGCI members are not trying to implement a far-reaching overhaul of the oil and gas sector, but rather are seeking to adapt it to a low-carbon future.

The OGCI as a potential climate leader

From this exploration of the extent and conditions under which oil and gas companies act as climate leaders or pioneers through four categories – positions, conditions, styles and strategies – the OGCI appears as an ambitious initiative with a significant capacity for leadership. We can categorize it as a pusher in terms of its position, for it exhibits both internal and external ambitions by investing resources and high-level personnel and establishing partnerships with state and non-state actors across the climate regime.

Table 3. The OGCI and its types of leadership.

Position	Type of leadership			
	Structural	Entrepreneurial	Cognitive	Exemplary
Pusher A/B	(x)	X	X	-

Even though there are few indications as to whether the OGCI will indeed carry through with its stated ambitions, the very fact that it is seeking to influence internal and external actors suggests that it would like to take on a leadership position (rather than merely act as a pioneer, which Liefferink and Wurzel (2017) defined as an actor who does not want to attract followers). Several factors in the context surrounding the OGCI would appear to support this idea: a growing space is opening for non-state actors to become engaged in climate action; increasingly specific targets – such as in the Paris Agreement – that have the potential to constrain actors and generate societal boundaries; and exponential growth in support for a low-carbon future, together with major advances in the required technology.

In considering both positions and conditions, we conclude that the OGCI is a bona fide attempt by the oil and gas industry to become engaged in climate governance, although on its own terms. For the moment, however, it is impossible to predict the outcome of this initiative.

As for the styles and strategies employed by the OGCI to enact leadership, these paint a picture of the initiative as being able to wield influence in a number of ways. Through its considerable resources, the OGCI and its firms can make use of structural leadership, a form usually only available to states. It also employs entrepreneurial and cognitive leadership types by establishing itself as the first collective effort of the oil and gas industry to address climate change, and by putting forward an initial vision of how the oil and gas sector might be integrated into a low-carbon world. However, as the analysis of its leadership styles shows, the OGCI does not have a radical vocation and is primarily designed to push for incremental changes, thereby allowing the industry to adapt to a non-fossil energy future.

Conclusion

This exploration of the positions, conditions, styles and strategies employed by a specific instance of the oil and gas sector, the OGCI, shows it is taking on a leadership role (a 'pusher' in the language of Liefferink and Wurzel 2017) in the governance of climate change. Although we are unaccustomed to thinking of oil and gas firms as taking on a more progressive role vis-à-vis climate change, this analysis of the OGCI would seem to indicate that the era of the climate-sceptic fossil fuel regime is at an end and that the industry is now attempting to sketch out a role for itself in a low-carbon world. But this attempt at claiming climate leadership on the part of the industry remains a new turn and, crucially, one that requires a more nuanced understanding of a sector that is increasingly marked by internal divisions in terms of policy, investment decisions and wider strategies.

An important line of questioning in this regard would be how the OGCI and other actors in the oil and gas sector implement leadership. For the moment, actions appear largely discursive (e.g. through statements) and network-based (e.g. by linking to high-level nodes within the climate regime), although more recently, concrete investments for low-carbon technologies have been announced.

Three more questions follow from the answers given to our initial analytical queries concerning the positions, motivations and strategies of the OGCI towards climate governance. First comes the question of why this is important. Although we could dismiss the OGCI as an as yet fruitless initiative, no immediate reasons have been found to discount it altogether. The value of such an early-stage study lies in showing how this initiative and its firms are trying to involve themselves in climate governance, and, in a post-Paris world, to redefine their role vis-à-vis an accelerating energy transition. Given the significant political, financial and regulatory power possessed by these firms, it is essential to track their current attempts to forge state and non-state partnerships throughout the climate regime.

The second question flowing from this is of a more conceptual nature and concerns the limitations of the employed approach, in particular the application of the Liefferink and Wurzel (2017) framework. The main one relates to the difficulty in assessing the position of actors along two purely linear dimensions – their internal and external environmental ambitions – for it is not entirely clear what the criteria or methods should be for attributing such characteristics to actors. The fuzziness of the internal-external divide could then be addressed by putting in place explicit definitions that are equally applicable to state and non-state actors. In the case of the oil and gas industry, 'internal' might refer to the sector rather than firm level, whereas 'external' might refer to ambitions directed at other sets of actors.

A related issue is the assumption of a single coherent position. Actors may well take on multiple positions simultaneously, as when oil and gas firms make climate commitments *and* expand oil production in short succession. A final conceptual issue concerns the notion that strategies (i.e. types and styles of leadership) are stable. This contribution has highlighted the co-evolution of external and internal dynamics, which implies that – especially for private actors who have fewer institutional constraints – it is unlikely single or even consistent strategies will be employed. It may therefore be more appropriate to focus on degrees of engagement rather than on given roles such as 'leader' and 'pioneer'.

The third and final question emerging from this discussion relates to its implications for polycentric climate governance. Caution is key here, as there are few guarantees that a nascent initiative will bear fruit and that its ambitions will remain unchanged. We can find a vivid illustration in BP's 'Beyond Petroleum' branding campaign, which BP eventually cast aside in favour of a renewed focus on its core business of oil and gas (Driessen 2003). Yet, there is cause to believe that something beyond a PR change may be occurring. For example, Total recently invested heavily in renewable technologies, for instance, buying a battery maker (Rascouet et al. 2016). Overall, however, the situation remains murky: ExxonMobil orchestrated a major green marketing campaign during the 2016 Rio Olympics, aiming to highlight its environmental ambitions and gloss over its efforts to derail climate science and policy (Henn 2016). Difficult as it may be to tease out the implications, I would nevertheless suggest three – from negative to positive. A rather pessimistic implication from the engagement of oil and gas firms in climate governance would be that a classic case of regulatory capture is at hand (Warren 2016). From a somewhat more neutral perspective, we could argue that multiple pathways are being developed towards a low-carbon world and that competition between them could lead to an upward ratchet in standards and practices (Braithwaite and Drahos 2000). On the optimistic edge of this spectrum of implications, it might be said that the engagement of the oil and gas sector in climate governance heralds the injection of vast and sorely needed resources – whether financial, technological or political – into efforts to stem climate change.

If the OGCI is truly to take on a leadership position in relation to the governance of our changing climate, we would do well to closely monitor how it translates its commitments into actions and hold it accountable by denouncing its shortcomings and lauding its successes. It also seems important to delve deeply into the low-carbon visions being put forward by oil and gas firms, which may be unrealistic and reliant on technological fixes rather than on a genuine desire to adapt. Last but certainly not least, we should keep a close eye on the degree of (non-) alignment between state and non-state approaches to climate change: a lesser gap between the two could indicate a greater potential for effective climate action.

Notes

1. Such as the Energy Transitions Commission (ETC), a high-level group of CEOs, civil society leaders, and former politicians that advises governments on climate mitigation and adaptation; or the World Business Council for Sustainable Developments (WBCSD) – led Low Carbon Technology Partnerships initiative (LCTPi).
2. National oil companies (NOCs) are state-led and include Saudi Aramco (Saudi Arabia), CNPC (China), Pemex (Mexico) and Statoil (Norway). By contrast, international oil companies (IOCs), such as BP, Shell and Total, are private-sector corporations. Though all face similar challenges, their agendas and relationship to regulation can differ significantly.
3. One should nonetheless bear in mind that CEOs are not the only decision-makers in these large firms. Decisions tend to taken with the broader executive board, supervisory boards and in some cases also with key investors.

4. In particular, three of the main trade organizations: the American Petroleum Institute (API), Fuels Europe and IPIECA. Each targets specific polities, gathers diverse memberships, and pursues different issues.

Acknowledgments

I have greatly benefited from the comments and patience of the Special Issue editors – Rüdiger Wurzel, Duncan Liefferink, and Diarmuid Torney – both during and after the INOGOV-funded workshop *Pioneers and Leaders in Polycentric Climate Governance*, which took place in Hull, UK on 15–16 September 2016. I would also like to thank the anonymous reviewers for their insightful comments, as well as all interviewees for their trust and time.

Disclosure statement

No potential conflict of interest was reported by the author.

ORCID

Matthew Bach http://orcid.org/0000-0003-1982-9671

References

Bäckstrand, K., et al., 2017. Non-state actors in global climate governance: from Copenhagen to Paris and beyond. *Environmental Politics*, 26 (4), 561–579. doi:10.1080/09644016.2017.1327485.

Banerjee, N., Song, L., and Hasemyer, D. 2015a. *Exxon believed deep dive into climate research would protect its business* [online]. Inside Climate News, 17 September 2016. Available from: https://insideclimatenews.org/news/16092015/exxon-believed-deep-dive-into-climate-research-would-protect-its-business [Accessed: 23 July 2016].

Banerjee, N., Song, L., and Hasemyer, D. 2015b. *Exxon's own research confirmed fossil fuels' role in global warming decades ago* [online]. Inside Climate News, 16 September 2016. Available from: https://insideclimatenews.org/news/15092015/Exxons-own-research-confirmed-fossil-fuels-role-in-global-warming [Accessed: 23 July 2016].

Bousso, R. and Schaps, K., 7 October 2015. *Oil bosses to meet in latest climate change offensive*. London: Reuters.

Braithwaite, J. and Drahos, P., 2000. *Global business regulation*. Cambridge: Cambridge University Press.

Brown, L.R., 2000. *The rise and fall of the global climate coalition* [online]. Earth Policy Institute, Rutgers University. Available from: http://www.earth-policy.org/plan_b_updates/2000/alert6 [Accessed: 23 July 2016].

Chan, S., et al. 2015. Reinvigorating international climate policy: a comprehensive framework for effective nonstate action. *Global Policy*, 6 (4), 466–473. doi:10.1111/1758-5899.12294.

Cherp, A., Jewell, J., and Goldthau, A., 2011. Governing global energy: systems, transitions, complexity. *Global Policy*, 2 (1), 75–88. doi:10.1111/gpol.2010.2.issue-1.

DDPP, 2015. Pathways to decarbonization. *Deep Decarbonization Pathways Project Report*. Paris: IDDRI and SDSN.

Downie, C., 2014. Transnational actors in environmental politics: strategies and influence in long negotiations. *Environmental Politics*, 23 (3), 376–394. doi:10.1080/09644016.2013.875252.

Driessen, P.K., 2003. BP—back to petroleum. *A Quarterly Review of Politics and Public Affairs*, 55 (1), 13–14.

Dudley, B. 2015.Why are oil and gas companies calling for more action on climate change? *Japan Today*, 14 Nov.

Eikeland, P.O. and Skjærseth, J.B., 2019. Oil and power industries' responses to EU emissions trading: laggards or low-carbon leaders?. *Environmental Politics*, 28 (1).

Froley, A., 2016. Engie launches 3-yr transition plan. *Natural Gas Europe*, 26 February.

Griffin, P.A., et al., 2015. Science and the stock market: investors' recognition of unburnable carbon. *Energy Economics*, 52, 1–12. doi:10.1016/j.eneco.2015.08.028.

Henn, J., 2016. ExxonMobil takes the Olympic gold in deceitful advertising. *Huffington Post*, 9 Jul.

Holland, R.A., et al., 2016. Bridging the gap between energy and the environment. *Energy Policy*, 92, 181–189. doi:10.1016/j.enpol.2016.01.037.

IPIECA, API and IOGP, 2011. *Petroleum industry guidelines for reporting greenhouse gas emissions*. Second edition ed. London: The Global Oil and Gas Industry Association for Environmental and Social Issues, The American Petroleum Institute, and International Association of Oil & Gas Producers.

Jennings, K., Grandoni, D., and Rust, S., 2015. How Exxon went from leader to skeptic on climate change research [online]. *Los Angeles Times*, 23 October 2015. Available from: http://graphics.latimes.com/exxon-research/ [Accessed 23 July 2016].

Jerving, S. et al., 2015. What Exxon knew about the earth's melting Arctic [online]. *Los Angeles Times*, 9 October 2015. Available from: http://graphics.latimes.com/exxon-arctic/ [Accessed 23 July 2016].

Levy, D. L., and Kolk, A., 2002. Strategic responses to global climate change: Conflicting pressures on multinationals in the oil industry. *Business and Politics*, 4 (3), 275-300. doi: 10.2202/1469-3569.1042

Liebermann, A. and Rust, S., 2015. Big oil braced for global warming while it fought regulations [online]. *Los Angeles Times*, 31 December 2015. Available from: http://graphics.latimes.com/oil-operations/ [Accessed 23 July 2016].

Liefferink, D. and Wurzel, R.K.W., 2017. Environmental leaders and pioneers: agents of change? *Journal of European Public Policy*, 24 (7), 951–968. doi:10.1080/13501763.2016.1161657.

Mathieu, C., 2015. *Carbon risk and the fossil fuel industry*. Paris: Actuelles de l'Ifri.

Mitchell, J.V., Marcel, V., and Mitchell, B., 2015. Oil and gas mismatches: finance, investment and climate policy. *Chatham House Research Paper, July 2015*. London: Chatham House.(30)

Nasiritousi, N., 2017. Fossil fuel emitters and climate change: unpacking the governance activities of large oil and gas companies. *Environmental Politics*, 26 (4), 621–647. doi:10.1080/09644016.2017.1320832.

OGCI, 2015. *More energy, lower emissions: catalyzing practical action on climate change. A report from the oil and gas climate initiative* [online]. Available from: http://www.oilandgasclimateinitiative.com/~/media/Files/O/Ogci/documents/ogci-report-2015-news.pdf [Accessed 23 July 2016].

Papaefthymiou, G. and Dragoon, K., 2016. Towards 100% renewable energy systems: uncapping power system flexibility. *Energy Policy*, 92, 69–82. doi:10.1016/j.enpol.2016.01.025.

Pattberg, P., 2016. *Environmental governance in the anthropocene: complexity, fragmentation and the role of transnational institutions* [online]. Inaugural address, Amsterdam, The Netherlands. Available from: https://research.vu.nl/ws/portalfiles/portal/1436301 [Accessed 23 July 2016].

Pattberg, P. and Widerberg, O., 2016. Transnational multistakeholder partnerships for sustainable development: conditions for success. *Ambio*, 45 (1), 42–51. doi:10.1007/s13280-015-0684-2.

Politiques énergétiques, 2015a. Video report on the 16 October 2015 CEO event [online]. Available from: https://www.politiques-energetiques.com/logci-paris-quand-les-petroliers-se-mobilisent-pour-le-climat [Accessed 21 August 2016].

Politiques énergétiques, 2015b. Video of the 16 October 2015 CEO press conference [online]. Available from: https://www.politiques-energetiques.com/logci-paris-quand-les-petroliers-se-mobilisent-pour-le-climat [Accessed 21 August 2016].

Rascouet, A., De Beaupuy, F., and Hirtenstein, A., 2016. Total to buy battery maker Saft in push to expand clean energy. *Bloomberg*, 9 May.

Revkin, A.C., 2009. Industry ignored its scientists on climate [online]. *New York Times*, 23 April. Available from: http://nyti.ms/1LCJW0Z [Accessed 23 August 2016].

RFF, 2016. *RFF's decision to divest* [online]. Rockefeller Family Fund. Available from: http://www.rffund.org/divestment [Accessed 21 August 2016].

Sampathkumar, M., 2016. *Oil and gas companies pledge $1 billion to reduce emissions — but they won't give up on fossil fuels* [online]. ThinkProgress on Medium, 17 November 2016. Available from: https://thinkprogress.org/1-billion-oil-gas-pledge-clean-energy-9f18f0ecb5dd#.tfrwa1d5a [Accessed 18 January 2017].

Solsvik, T. and Fouche, G., 2016. Big Oil pledges $1 billion for gas technologies to fight climate change. *Reuters*, 4 November.

Stevens, P., 2016. International oil companies, the death of the old business model. *Chatham House Research Papers, May 2016*. London: Chatham House.

Timperley, J., 2016. E.ON completes split of fossil fuel and renewable operations. *The Guardian*, 4 Jan

UN, 2014. *Press release: industry leaders, including energy companies, forge partnerships to advance climate solutions and reduce short-lived climate pollutants* [online]. Available from: http://www.un.org/climatechange/summit/wp-content/uploads/sites/2/2014/05/INDUSTRY-PR.pdf [Accessed 21 August 2016].

UNEP, 2015. *Climate commitments of subnational actors and business. a quantitative assessment of their emission reduction impact*. Nairobi: United National Environment Programme.

van der Ploeg, F., 2016. Fossil fuel producers under threat. *Oxford Review of Economic Policy*, 32 (2), 206–222. doi:10.1093/oxrep/grw004.

Verbong, G. and Loorbach, D., 2012. *Governing the energy transition: reality, illusion or necessity?* London: Routledge.

Warren, E., 2016. *Corporate capture of the rulemaking process* [online]. RegBlog, 14 June 2016. Available from: https://www.theregreview.org/2016/06/14/warren-corporate-capture-of-the-rulemaking-process/ [Accessed 21 August 2016].

Wurzel, R., Liefferink, D., and Torney, D., 2019. Pioneers, leaders and followers in multilevel and polycentric climate governance. *Environmental Politics*, 28 (1).

Oil and power industries' responses to EU emissions trading: laggards or low-carbon leaders?

Per Ove Eikeland and Jon Birger Skjærseth

ABSTRACT
How have petroleum and power companies and their European industry associations responded to the EU emissions trading system (ETS)? Responses can be political, directed externally towards the initiation and reforms of the EU ETS itself, or internally and market-based, directed at low-carbon solutions. Proactive response strategies shape companies' leadership potential. Variation in responses is explained by two models that differ in assumptions about corporate behaviour as well as the wider multilevel regulatory context in which companies operate. Responses are found to have converged within the two industries, with reactive companies following the proactive ones. Secondly, responses between the two industries increasingly diverge, with the power industry becoming much more proactive than the petroleum industry. The main explanation is found in the differing relevance of the two models and the wider regulatory context, particularly differing exposure to international competition and weak international climate agreements.

Introduction

Energy industries are pivotal to achieve decarbonisation of energy supply because they are a major source of the climate change problem and because their entrepreneurship is called for in developing solutions. In 2005, the EU launched its climate policy flagship – the emissions trading system (EU ETS) – to spur more ambitious corporate climate strategies in these and other industries. Since then, the EU has reformed the ETS in several rounds. We analyse strategic responses of individual companies and associations representing petroleum and electric power supply industries to the evolving ETS. Based in the corporate strategy literature, we distinguish between 'proactive' and 'reactive' corporate strategies emerging from political responses to the introduction and reforms of the EU ETS, and adaptation in the market through action to reduce carbon emissions in the short and long term.

Major petroleum and power companies are important agents of change (Bach 2019 – this volume). Such actors have the capacity to act as pioneers 'ahead of the troops' or leaders that seek to attract followers and to exercise different types of leadership/pioneership (Liefferink and Wurzel 2017, Skjærseth 2017). Our contribution links the literature on corporate strategies to leadership by relating political responses to external leadership ambitions and market adaptation to internal leadership ambitions (Liefferink and Wurzel 2017). Actual leadership takes place if laggard companies follow the leaders to bring about more ambitious collective industry-association actions and positions. To what extent and how did corporate strategies emerge and consolidate under the evolving EU ETS, from the system's initiation until the recent reforms for 2030? Under what conditions did the EU ETS affect the strategies and leadership of the energy companies, individually and collectively?

We explain variation in response strategies based on literature on the relationship between regulation and corporate strategies (Skjærseth and Eikeland 2013). We develop two 'models' based on different behavioural assumptions of how we expect companies to respond to the EU ETS. However, the EU ETS is obviously not the only factor shaping corporate strategies. To further explain and identify conditions for corporate responses, we analyse the EU ETS as a governance system related to other climate regulations (Homsy and Warner 2015). Our contribution thus also informs the debate on whether different mixes of multilevel and polycentric governance facilitate corporate leadership (Liefferink and Wurzel 2018). These 'top-down' and 'bottom-up' governance types resonate with two key EU ETS features that shape companies' room for exercising leadership. On the one hand, the ETS is an EU-level, harmonised and mandatory cap-and-trade system for the installations/companies operating within the system. On the other hand, ETS companies have significant independence and autonomy to shape their short- and long-term strategies within the system and their respective industry branches. An underlying theme in this contribution is whether this balance between 'bottom-up' polycentric and 'top-down' multilevel types of governance is 'right' for leadership to emerge. We expect leadership to emerge to the extent that governance mixes trigger continuous proactive change while simultaneously providing sufficient room for independent adaptation and responses across different industries and companies.

We contribute on three fronts to the literature on the relationship between emissions trading and corporate climate strategies within a wider governance approach (e.g. Ellerman *et al.* 2010, Meckling 2011, Skjærseth and Eikeland 2013). First, we offer a new 'model'-based approach to studying this relationship grounded in various assumptions of corporate behaviour. Second, we link the corporate strategy literature to leadership and types of governance. Third, we contribute empirically by comparing corporate responses to the EU ETS in the petroleum and electric power industries. To our knowledge, such studies have been lacking.

For the petroleum and electric power sectors, we selected two companies within each sector – ExxonMobil and Shell, and Vattenfall and RWE, respectively – according to two criteria. First, differences in climate strategies prior to the EU ETS, to shed light on the regulatory conditions under which different strategies emerge and change. Second, they are major players in the oil and electricity markets, which give them leadership potential within European industry associations for the electric power and petroleum sectors – Eurelectric and Europia/FuelsEurope.

We build on multiple sources, including companies' self-reporting, secondary information and interviews with representatives from companies and industry associations.[1] We list the interviewees at the end.

Analytical point of departure

We develop two alternative 'models' that will generate different expectations about 'reactive' and 'proactive' corporate response strategies to the EU ETS based on different rationality assumptions about corporate behaviour.[2] These strategies are ideal-typical opposite poles, and we cannot expect real-life companies engaged in a wide range of activities to fit perfectly with such opposite extremes. The aim is to assess the degree of fit between expectations and observations in the content and direction of corporate strategies from before the EU ETS was adopted. *Political response* indicates whether companies support more stringent regulation, or actively resist and oppose regulation. *Market responses* refer to compliance measures; these include carbon-abatement measures, trading to compensate for abatement and measures to spur innovation in long-term low-carbon solutions (Kolk and Pinkse 2004, 2008).[3] Corporate response strategies relate to leadership/pioneership through different degrees of internal and external ambitions (Liefferink and Wurzel 2017). High external (political) support and internal (market) ambitions allow companies to become a leader. Conversely, low internal and external ambitions make companies potential laggards.

'Reactive' corporate responses: potential laggards

This model (Table 1) sees the firm as a unitary rational profit-maximising agent that adopts its strategies on the basis of full information of the relative costs of various alternatives (Gravelle and Rees 1981, Ambec et al. 2011). The model is static: prior to regulation, companies would have adapted optimally in the output and input markets at levels reflecting marginal income equalling marginal costs. Any new environmental regulation, like the EU ETS, would impose net costs on the company, eroding profits and competitiveness, unless all competitors are subject to similar regulatory costs.[4] The EU ETS will thus appear as a regulatory threat (Bohr 2016). Politically, we expect that companies will oppose the system. Observed opposition expressed in position papers to EU ETS consultations will be in line with this expectation.[5]

As for market responses, companies will comply by adopting only low-cost incremental business-as-usual abatement options. They will base the actual choice of options on cost-ranking, in line with the least-cost compliance principle.[6] Contrary behaviour would be illogical within this model where full-information profit-maximising companies have already discovered all the 'low-hanging fruits' and taken advantage of those opportunities before regulation was put into place (Ambec et al. 2011). We expect business-as-usual activities in the short term, and weak focus on new low-carbon innovation for the longer term. Observation of only short-term compliance measures and minor engagement in long-term low-carbon solutions and R&D will support this model.

Table 1. 'Reactive' corporate response model.

Key external explanatory factor	Behavioural assumption	Wider economic, social and political context	Expected response to ETS
Mandatory regulation	Rationality and profit maximisation: minimisation of new regulatory costs	Irrelevant for explaining (change in) strategic decisions	Reactive strategy: political opposition and only low-cost market adaption. Potential laggard.

'Proactive' corporate responses: potential leaders

This alternative model (Table 2) sees the firm as only boundedly rational (Cyert and March 1963). Companies strive for profits, but are unable to make optimal choices – because of market failures, organisational inertia, and managers being constrained in information and in cognitive capacity for making explicit and timely calculations of optimality (Cyert and March 1963, Simon 1976). Instead, companies base their decisions on sequential attention, risk averseness, and standard operating procedures, habits and routines (Cyert and March 1963).

Based on such assumptions, Porter and van der Linde (1995) hypothesised a different view of the relationship between regulation and corporate strategies. They suggested that appropriate environmental regulation could generate new attention of companies to earlier non-apprehended opportunities, spur learning about resource inefficiencies and technological improvements, reduce uncertainty about future investments, create pressures to motivate innovation offsetting compliance costs, and in fact strengthen the international competitiveness of regulated companies. This model would predict political support for the ETS, since companies should rapidly discover and focus on new business opportunities. Dynamically, we expect the discovery of new opportunities to increase political support over time for a stringent EU ETS. Support for an increasingly stringent ETS expressed in position papers will be in line with this expectation.

As to market response, the model predicts that companies will start searching for new market opportunities beyond business-as-usual, in order to create early-mover advantages. We can expect incremental innovation (short-term abatement measures beyond what is needed for compliance) and long-term R&D directed at *new* large-scale innovation, since company management has re-directed attention towards opportunities previously unheeded. Observations of beyond compliance abatement measures and upscaling of engagement in new long term low carbon solutions and R&D investments will fit proactive market responses.

Table 2. 'Proactive' corporate response model.

Key external explanatory factors	Behavioural assumption	Wider economic, social and political context	Expected response to ETS
Mandatory regulation	Bounded rationality: myopic attention and search for new market opportunities	Dynamic competition relevant for explaining (change in) strategic decisions	Proactive strategy: support of increasingly stringent regulation and new low carbon market opportunities found and acted upon. Potential leader.

Factors conditioning corporate 'proactive' responses

The models above predict similar responses by all companies that the EU ETS regulates. In a real-world situation, however, both company-internal and external factors may condition such responses. Space constraints do not permit full empirical examination here of company-internal factors, such as variation in organisational structures, management and capability to act on changes in the external environment (Teece 2007, Kolk and Pinkse 2008).

Moreover, we limit our discussion to the 'proactive' response model that corresponds with potential leadership, specifically addressed in this volume.

Firstly, companies and sectors differed in carbon-intensities of technologies inherited from before the EU ETS was adopted, entailing that the regulatory costs from the EU ETS would differ between companies. The EU ETS would accordingly be more stringent for carbon-intensive companies. Porter and van der Linde (1995) added 'stringency' of the regulation as a conditioning factor for spurring companies to start focusing on long-term learning and innovation. We would expect to see more 'proactive' responses in companies facing the highest regulatory costs from the EU ETS: highly carbon-intensive companies that faced allowance deficits and had to pay for allowances.

Secondly, the regulatory risks of the ETS could differ across companies and sectors, due to differences in international trade- and competition-intensities. Porter and van der Linde (1995) formulated such regulatory risks as other conditioning factors for environmental regulations to spur the search for early-mover advantages and innovation – implying that a regulation should be developed in line with or just slightly ahead of other countries. This means that we will not expect trade-intensive companies exposed to competition from outside the EU ETS area to respond proactively unless they expect other countries to follow up with similarly stringent climate regulation. By contrast, we can expect 'proactive' responses from companies that are not exposed to international competition. This factor directs attention to the alignment between the EU ETS and international climate agreements.

Porter and van der Linde (1995) also added another conditioning factor for a regulation to spur competitive advantage through innovation: that it should be aligned with other regulatory measures so as to create synergy rather than confusing regulatory signals. Variation in response between sectors and companies may be expected, depending on whether other sector-specific EU and national regulations send regulatory signals consistent with those of the EU ETS.

Summing up, we expect the EU ETS to shape proactive strategies when carbon-intensive companies can be depicted as boundedly rational actors and are exposed to a regulatory framework that is stringent, aligned with other relevant policies at EU and national levels, and promotes a level playing field internationally. Proactive strategies characterised by high external (political) and internal (market) ambitions may result in actual leadership to the extent that laggards follow, and allow more ambitious collective positions and action by industry associations. We based these expectations related to the 'proactive' model on the assumption that the regulatory design of different governance types matters for leadership to emerge.

Responses to the EU ETS

EU ETS reforms

We explore individual corporate and collective industry responses in three ETS phases demarcated by reforms to the system: initial responses to the establishment of the ETS; responses to the 2008 revision; early responses to the revision initiated in 2015.

From its inception in the late 1990s, the EU ETS has evolved with the formal adoption of the EU ETS Directive in 2003 and launch of the carbon market in 2005 (Skjærseth and Wettestad 2008). The EU significantly revised the system in 2008 for 2013–2020, and from 2015 for 2021–2030 (Skjærseth and Wettestad 2010). A contested design issue was how to set the cap that determined the stringency of the system in terms of emissions reduction. Initially, member states demanded national control over allocation. In 2008 they accepted centralised allocation, with the EU-level cap set to ensure reduction in emissions by 2020. Also contested through all phases was the method of allocation: free allowances, or auctioning. Member states gradually accepted more auctioning for the period 2013–2020, with mandatory auctioning for power companies as the main rule and a gradual phase-in for energy-intensive industries exposed to international competition, including petroleum companies.[7] Reforms from 2015 for 2021–2030 have also triggered a conflict regarding how to counter the considerable build-up of surplus allowances (European Commission 2015).

The reformed ETS from 2008 formed part of a larger EU climate and energy package that included binding policies on sectors not covered by the ETS, renewables, carbon capture and storage (CCS), energy efficiency, fuel quality and car emissions (Skjærseth *et al.* 2016). Together, these policies aim at reducing emissions and increasing energy efficiency and the share of renewables in primary energy consumption by 20% by 2020. For 2030, emissions within the EU are to be reduced by at least 40% compared to 1990 levels. The long-term EU vision is an 80–95% reduction of emissions by 2050.

Initial corporate responses

Initial responses to the EU ETS varied significantly between companies in the power and petroleum industries. Within the electric power industry, Vattenfall, which had mainly operated non-fossil energy plants in its home market Sweden (hydropower and nuclear), became a major operator of coal power on the continent. RWE, which originated as the main operator of coal-based power in Germany, became more diversified, particularly after acquisitions of natural gas. Although the two companies became more similar in their carbon-intensities in the market, clear differences existed in how they viewed the climate change problem and their strategic role in dealing with it. *Time* magazine hailed Vattenfall's CEO for his leading role in spurring global political action (Eikeland 2013, p. 5–7). In 2001, Vattenfall became the first power company in Europe to adopt a long-term strategy for developing CCS to make its coal-powered plants carbon-neutral. RWE pursued a far more reluctant climate strategy, acknowledging the climate problem but emphasising the scientific uncertainties (Eikeland 2013, p. 61). The variation between the two companies became evident in their differing political *responses* to the emerging EU ETS: Vattenfall proactively supported the initiation of the ETS, whereas RWE was strongly opposed (Eikeland 2013, RWE 2001, p. 31).

Petroleum companies with significant and comparable operations in Europe also displayed highly differing climate strategies when the EU ETS was initiated (Skjærseth 2013). Shell acknowledged the problem and supported the Kyoto Protocol; it had pioneered company internal emissions trading and adopted GHG emissions targets and measures with significant long-term implications for mode of operation and business orientation. Conversely, ExxonMobil saw the problem merely as a 'legitimate concern'; it opposed the Kyoto Protocol and had not adopted GHG emissions targets and measures (Skjærseth 2003, Skjærseth and Skodvin 2003). This variation continued with highly differing political responses to the initiation of the EU ETS. Shell proactively supported the idea, whereas ExxonMobil lobbied actively against the initiation of the EU ETS (Skjærseth 2013).

Significant differences appeared also in the political responses from European industry associations. Europia (now FuelsEurope) accepted the EU ETS only reluctantly, as a non-mandatory pilot for a larger global system (Skjærseth 2013). Eurelectric's response evolved from early scepticism to becoming largely positive towards a mandatory system (Eurelectric 2002, Eikeland 2013). Eurelectric's support increased after association-initiated allowance trade simulations showed that the power industry could gain from the EU ETS spurring higher power prices. These results also dampened RWE's opposition to the EU ETS (RWE 2003, p. 23).

Once the EU launched the ETS in 2005, *market responses* varied, with electric power companies generally implementing abatement measures more far-reaching in scope and scale than petroleum companies and what they needed for short-term compliance (Eikeland and Skjærseth 2013). Both sectors implemented many typical no-regret solutions (energy efficiency), low-cost solutions (buying allowances) and measures demanded by the ETS (monitoring, reporting and verification) as well as establishing new trading desks. The companies in both sectors increased their R&D focus on developing and demonstrating CCS technology. For the electric power companies, this signified a *new* technology field. For petroleum companies, upscaled activities represented deeper involvement in a technology already familiar to them, with CO_2 captured and injected as compressed gas to increase production from mature oil fields.

Within both industries, we find a clear trend whereby companies converge in market responses, in the sense that those lagging behind in GHG-mitigation efforts caught up with the companies that had pioneered such efforts. In the power industry, we can see this with both RWE and Vattenfall increasing their investments in developing CCS technology, renewable energy and other low-carbon solutions year-on-year after the ETS started up (Eikeland 2013).

Shell and Exxon likewise demonstrated convergence in scaling up activities that had been part of their business since long before the EU ETS: energy-efficiency efforts and investments in CCS. Exxon in particular reported good achievements in energy efficiency, and set targets for reducing emissions from flaring. A notable change appeared: Exxon now took a softer position on climate policy. In 2007, Exxon acknowledged responsibility in helping to alleviate the climate problem and halted its funding of several anti-climate lobby groups. It also accepted carbon pricing by international taxation, but not cap-and-trade (Exxon 2007, Skjærseth 2013, p.112).

We find that the introduction of the first mandatory European climate regulation – the EU ETS – apparently did affect corporate strategies. The direction of strategic responses did initially fit well with 'reactive' laggards (Exxon, RWE) and 'proactive' potential leaders (Vattenfall, Shell) with higher internal and external ambitions. Strategies converged over time with the reactive catching up with the more proactive companies.

Responses to the 2008 reform

The 2008 EU ETS reforms triggered new political responses. In its position paper on the ETS revision proposal, Europia for the first time explicitly accepted the ETS in light of concerns about climate change (Europia 2008). Exxon still opposed emissions trading, but its softer stance on climate policy increased the leverage of pro-ETS companies like Shell to shape the association's position. This said, Shell and Exxon, as well as Europia, lobbied fiercely against more stringent regulation of the petroleum companies (Europia 2010a). The Commission proposed to treat oil refining on a par with other energy-supply activities and to subject it to full auctioning (European Commission 2006). The petroleum sector prevailed, gaining recognition as an 'energy-intensive sector' to be allocated free allowances based on benchmarks from 2013. By contrast, Vattenfall, RWE and Eurelectric now supported full auctioning of allowances for the electric power industry (Eurelectric 2007, Skjærseth and Eikeland 2013).

We must view the initial post-2008 *market* responses in light of carbon prices that rose to nearly EUR 30/tonne. In 2009, 61 CEOs of the major European electric power companies, including RWE and Vattenfall, signed an industry declaration on de-carbonising electricity supply in Europe by 2050 (Eurelectric 2009a). This followed a Eurelectric-initiated scenario study indicating that the costs of achieving carbon-neutral electricity supply would be nearly offset by saved allowance payments and revenues from higher electricity tariffs under the EU ETS (Eurelectric 2009b). Both Vattenfall and RWE continued converging in more proactive market responses: setting quantitative short- and long-term GHG emissions-reduction goals, increasing R&D in low-carbon solutions, greater pace in investments in renewables, signalling more investments in natural gas-based power and an end to new coal investments (Eikeland 2013, p. 65).

Beyond this, the petroleum companies adapted by refreshing R&D in low-carbon solutions such as CCS and advanced biofuels for transport, and brushing up on low-carbon energy future scenario planning. Shell continued preparing for an alternative long-term future by incorporating expected costs of CO_2 emissions into its financial planning of, and decisions on, major projects. Climate policy also gained a more prominent place in Exxon's Outlook for Energy, including expected carbon prices (ExxonMobil 2010, Coll 2011).

After carbon prices plunged from around 2009, we observe growing differences between the industries in market adaptation activities. However, one change is shared among all companies in both industries: the cancellation of planned large-scale CCS demonstration projects (Skjærseth *et al.* 2016).

The electric power industry experienced massive increases in investments in renewable energy fuelled by national subsidies: both Vattenfall and RWE carried out strategic shifts and major restructuring to accommodate stepped-up investments (Vattenfall 2015, Handelsblatt Global 2016). Most electric power companies, including Vattenfall and RWE, also signed a new Eurelectric-organised agreement promising no new coal plants in Europe after 2020 (Guardian 2017). For their part, petroleum companies have been painting a gloomy picture of future opportunities for the European refining industry because of the decreasing demand for petroleum products and reduced refinery margins (Europia 2010a, Skjærseth 2013). They perceived that a combination of factors caused this trend: costs of the EU ETS, tighter fuel specifications, restrictions on car emissions and support for non-fossil fuels.

New political initiatives for fixing the EU ETS to counter falling carbon prices came in 2012 with the Commission's proposal to postpone or 'backload' auctioning of 900 million allowances, spurring new political responses by companies and industry associations. Europia and Exxon now joined with a range of other associations representing energy-intensive industries in strong opposition, claiming this would increase the costs for the industries. Shell supported backloading, emphasising that higher carbon prices would increase the funding available from the set aside of auctioning revenues from the ETS specified for development of CCS and innovative renewable energy technologies (NER300 programme).

Eurelectric, unlike Europia, strongly supported backloading and other more permanent structural measures as necessary for the EU to signal commitment to a strong ETS (Eurelectric 2013). Vattenfall gave firm support to backloading, but RWE was reluctant, arguing that the measure was unnecessary because the cap would ensure GHG emissions reductions even without the price signal (RWE 2013, Vattenfall 2013).

Despite such company differences, we find a clear trend of growing difference in responses between the two industry associations. Eurelectric increasingly lobbied for making the ETS more stringent, in line with the preferences of Vattenfall and other leading member companies adapting most proactively in the market. Europia lobbied to keep the ETS lenient, compromising between the preferences of the 'leading' and 'lagging' member companies supporting and opposing the system.

Differences have also been reaffirmed recently in the negotiations on the revised EU ETS for 2021–2030. The main part of this reform has dealt with imbalances between supply and demand of allowances, with the aim of raising the carbon price. A more stringent cap and measures to withdraw surplus allowances from the carbon market have responded to this challenge.[8]

Political responses to these reforms by FuelsEurope (previously Europia) reaffirm that the oil industry operating in Europe accepts the ETS as the EU's main climate policy instrument, to be kept as lenient as possible (FuelsEurope 2014, 2015). ExxonMobil still opposes cap-and-trade, characterising it as unnecessarily costly, complex and ineffective (Tillerson 2010, CDP 2015). Nevertheless, Exxon supported the Paris Agreement and has urged President Trump not to withdraw US participation (Financial Times 2017).[9] In contrast, Vattenfall, RWE and Eurelectric have supported a stringent EU ETS for 2030 as a cornerstone of the EU's energy and climate policy (Eurelectric 2016). For example, Eurelectric acknowledges that the EU 40% target for 2030 is in the lower end for achieving decarbonisation by 2050 and has supported a stringent ETS cap and measures to deal with allowance surplus and bring carbon prices up (Eurelectric 2016).

The upshot is that the petroleum and power companies and their industry associations have responded increasingly divergently to ETS reforms. The petroleum industry has declared high EU ETS allowance prices based on a stringent system to be a threat to surviving in Europe. This indicates that actual leadership towards high climate ambitions has not emanated within Europia. Conversely, the electric power industry association has reached the opposite conclusion: higher ETS allowance prices based on a stringent system is a preferred option for securing remuneration of past investments and incentivising new low-carbon investments in the future. This indicates that actual leadership that laggards follow has occurred in the case of Eurelectric.

Summing up observed strategic responses

The first observation is that political and market responses, while mixed, have been converging between companies within the same industry, with 'reactive companies' catching up with the most 'proactive' ones to different degrees. The second observation is related to the increasingly diverging political responses from the industry associations. Europia came to accept the EU ETS but lobbied against reforms that would make the system more stringent, whereas Eurelectric has increasingly supported reforms that would make the system more stringent. In essence, corporate climate strategies within the sectors have become more similar, whereas differences between the petroleum and electric power industries have increased.

Explaining responses

Our analysis starts from two 'reactive' and 'proactive' responses models that depict corporate behaviour differently, thus providing alternative expectations for corporate responses to the EU ETS, and potential leadership. Some elements of the responses observed across all companies correspond well with what we could expect from companies portrayed by the first 'reactive' corporate responses model 'as minimisers of regulatory costs', already fully informed and optimally adapted. This includes cases where companies have opposed the EU ETS or specific design elements and have implemented typical low-cost and no-regret solutions, such as energy efficiency. Moreover, Exxon's opposition to cap-and-trade and Europia's sustained political reluctance towards making the system more stringent correspond to expectations from this model.

In contrast, we can better explain Eurelectric's proactive political response, and even more so, the power companies' dynamic development of external climate ambitions, towards supporting greater stringency in the EU ETS by the alternative 'proactive' corporate responses model. This model portrays companies as only 'boundedly rational', leaving them with the potential to discover new commercial opportunities from attention triggered by regulation, such as the EU ETS. This shows that the ETS may stimulate leadership characterised by high climate ambitions. The ETS spurred Eurelectric to conduct industrial joint trade simulations that generated collective learning about new income opportunities stemming from higher electricity prices (Skjærseth and Eikeland 2013).

After the ETS was up and running, we observe upscaling of R&D and long-term innovation efforts for low-carbon solutions particularly in the power industry, corresponding with expectations from the dynamic 'proactive responses' model. The Eurelectric-conducted joint scenario projects showing opportunities for industry expansion in a low-carbon future were important. Recognition that these opportunities could disappear if the ETS broke down triggered political acceptance of mandatory auctioning when the system was up for reform in 2008. This model thus explains also Eurelectric's growing support of a more stringent ETS. Eurelectric justified its support by citing the importance of a market signal to sustain attention to future low-carbon solutions.

Even the more temporary supportive position of ExxonMobil and Europia are explicable by the 'proactive responses' model, as major oil companies saw opportunities for developing low-carbon solutions like CCS, for which auctioning revenues were set aside under the 2008 reform.

Thus, we see that both 'cost minimisation' and 'new market opportunities' responses have shaped different *elements* of the companies' and industry associations strategies over time. This implies first that EU ETS has had mixed effects on pre-existing proactive and reactive corporate strategies through two different behavioural mechanisms. Second, it implies that the two behavioural models are insufficient to fully explain the increasingly divergent responses by the oil and power industries.

Factors conditioning proactive responses

We can better explain some additional variation in responses by accounting for factors at different governance levels, held to condition responses under the 'proactive response model'. We observed that, despite stimulating new attention and efforts towards developing long-term low-carbon solutions across industries and companies, we recorded the most extensive proactive learning and innovation responses in the electric power industry where internal and external climate ambitions increased.

Firstly, for an environmental policy instrument to trigger new attention and learning, and thus the emergence of climate leaders, it should be 'stringent'. All four companies were highly carbon intensive (coal, oil, gas) and potentially vulnerable to stringent climate regulation. However, the power industry became exposed to significantly more stringent regulation than the oil industry. After the ETS came about, most electric power companies emerged with a deficit of allowances compared to actual emissions, whereas petroleum companies came out roughly in balance (Skjærseth and Eikeland 2013). For the second trading period, the ETS was made more stringent for the electric power industry (partly auctioning) and even more so for the third trading period (full auctioning). In contrast, continued free allowances provided less stringent pressure on the petroleum industry. As expected, increasingly stringent regulation for the power companies correspond with increasingly 'proactive' strategies.

Secondly, adding regulatory market risk as a conditional factor could help further explain the variation in response between the petroleum and electric power companies and industries, and in particular the trend of diverging political responses between the industry trade associations. Some initial market risk (regulation slightly ahead of competitors) would be necessary to trigger attention and development of new opportunities in companies – but sustained differences in regulatory costs and market risks could also create competitive disadvantage, thus backfiring on strategic response. The electric power industry was not exposed to risk from international competition due to very low transmission capacity in and out of the EU area. The petroleum industry was more exposed, as trade in and out of the EU was growing because of the mismatch between production and consumption of petrol versus diesel.

The 2009 Copenhagen climate negotiations could have led to a reduction in such market risks, with other nations committing to more stringent climate policies, including carbon pricing. However, the failure of the Copenhagen talks and the establishment of a federal US cap-and-trade system may help to explain the opposition of the petroleum industry to making the ETS more stringent in later reforms. In fact, European energy-intensive industries reacted to Copenhagen by voicing fears that the EU could become a permanent lone-mover in climate policy (rather than a leader that would be followed by others), which would rob the industries of potential early-mover advantages. The extent to which the 2015 Paris Agreement will level the playing field remains to be seen. Difference in regulatory market risk has prevailed and it adds to our understanding of diverging responses. As this difference between the power and petroleum industries has remained roughly constant over time, low market risk has interacted with increasing 'stringency' and shaped more proactive strategies and leadership potential in the power industry.

A third conditioning factor was that environmental regulations should be co-ordinated at various governance levels to provide synergies with the EU ETS. Inconsistent regulations have affected all companies and both sectors. At a general level and in addition to failed international action, all four companies and the two industry associations are exposed – indirectly or directly – to roughly the same inconsistencies in EU climate- and energy policies. One example is that other EU energy and climate policies (energy efficiency and renewable energy policies specifically) contributed to depress already falling allowance prices, counteracting the intended effects of the EU ETS (Skjærseth and Eikeland 2013). Another example was carbon storage policies adopted at the national level that led petroleum and electric power companies alike to cancel planned large-scale demonstration projects. Such CCS-technology deployment problems are particularly unsettling for petroleum companies, since their core product – fossil fuels – makes CCS or equivalent technologies the only viable option currently available for decarbonisation. The electric power industry, by contrast, has more flexibility in decarbonisation, since its core product – electricity – is an energy carrier, not a specific primary fuel. These observations indicate that companies' flexibility to cope with inconsistencies appears more important than inconsistent regulation for explaining increasing divergence and leadership potential.

The latter point also draws attention to internal corporate factors, to which we have not given much attention in our contribution. We could better explain different response strategies within the same sectors, such as between Shell and ExxonMobil, by including management and capability to act on changes. Still, we can offer a cautious conclusion as to the conditions affecting corporate response strategies. The positive impact of regulation on proactive strategies and potential leadership is likely to increase when it is 'stringent' and not too far ahead of other countries, to avoid competitive disadvantages for companies exposed to international competition. Low regulatory market risk and high regulatory stringency seems to interact in promoting proactive strategies and leadership.

Conclusions

Our first question was directed at assessing change in corporate response strategies from when the ETS was initiated until the recent reforms for 2030. We examined how the response strategies of major energy companies evolved and aggregated as collective positions in industry associations. Within both the petroleum and electric power industries, our first conclusion is that political responses to the EU ETS converged as the regulation evolved. This was to varying degrees followed in the market by implementation of low-carbon solutions for the short and long term. In essence, the EU ETS as the first EU climate regulation to affect energy companies was instrumental in getting the more 'reactive' companies to follow the more 'proactive' ones in support of regulation and adaptation in the market.

Our second conclusion concerns different and increasingly divergent responses to the evolving ETS between the petroleum and electric power industries. Both industry associations accepted the initiation of the EU ETS, with Eurelectric more supportive than Europia. Europia maintained its acceptance of the regulation but consistently opposed making it more stringent, followed by growing concerns about competitiveness and carbon leakage. We have seen how the power companies strengthened their internal and external climate ambitions and that Eurelectric became increasingly supportive of a stringent EU ETS, followed by radical market ambitions aimed at decarbonising power supply in Europe. This observation indicates that the ETS helped spur actual leadership within the power industry by which the laggards followed, here illustrated by RWE and Vattenfall.

Our second question aimed at explaining variation in responses. We began with two 'models' of corporate behaviour. Here our main conclusion is that both models – the 'reactive' response model based on minimisation of regulatory costs and the 'proactive' response model based on short-sighted myopic attention and search for new market opportunities – explain different elements of corporate responses in different phases. However, the dynamic 'proactive' response model best represents how the ETS gradually triggered new lines of attention, learning and innovation in the electric power industry.

The extent to which the EU ETS triggered proactive responses and potential leadership has been conditioned by the wider set of regulations at different levels of governance. As to 'stringency' connected to the instrument itself, the ETS still deviates from this condition, although the 2008 revision brought the electric power industry closer to such an ideal regulatory situation, corresponding with the increase in proactive behaviour. This shows that cap-and-trade has significant potential to spur leadership when it is sufficiently stringent. The EU has not achieved full alignment between policies at the EU and national levels, as seen in national regulation of CCS deployment. Finally, the failure of the 2009 Copenhagen climate conference to get other countries to commit to set GHG emissions reduction goals led European industries to ask whether the EU ETS would bring permanent regulatory disadvantages rather than early-mover advantages. Differing exposure to international competition in the absence of international climate regulation is probably the most important factor that has conditioned differing responses in the petroleum and electric power industries. Unlike the petroleum industry, the electric power industry is not exposed to competition beyond Europe. The Paris Agreement *may* level the playing field for the petroleum industry in the long run. ExxonMobil's support of Paris appears promising in this regard.

The focus on the dynamic interaction between the EU ETS and corporate response strategies has also taken us a step towards understanding how climate leadership may be realised through the ability of industry-level associations to spur collective learning about opportunities and to decide on collective response strategies. There are various avenues for improving knowledge about corporate strategies and leadership. In terms of theory, our approach could be expanded to explore internal company factors more systematically and how individual and societal norms may affect corporate strategies. Empirically, more studies of corporate actors in different industries are needed, because private-sector companies will be the key transformation agents in dealing with the long-term challenge of climate change. In particular, we need a better understanding of how industry-level associations aggregate and integrate corporate strategies, and their roles in exercising leadership.

The conclusions indicate first that the 'proactive' response model – emphasising the 'stringency' of public regulation – is mostly compatible with multilevel governance that assumes a strong role for governing authorities. Second, the balance between 'top-down' multilevel regulation and 'bottom-up' polycentric autonomy also seems to matter for companies' strategies and leadership to emerge. Specifically, our contribution shows how the EU ETS has enabled leadership in the power industry. In this industry, increasing 'stringency' imposed from the 'top' combined with wide room for companies to choose their own response strategies have triggered proactive strategies and leadership among many corporate actors. Finally, the study has shown that the effect of governance systems on corporate strategies and leadership is conditioned by specific individual and collective actor characteristics. Low exposure to international competition and high flexibility to decarbonise has interacted positively with increasing regulatory 'stringency' and independence for companies to shape their own response strategies. This interaction led to increasingly proactive strategies and leadership in the power industry.

Notes

1. We have used interviews as background information.
2. The EU ETS is a mandatory system that will entail costs for all installations/companies included compared to a situation without the EU ETS. We therefore rule out 'indifference' and assume that all actors will respond in one way or another.
3. A more drastic response would be relocating production to other countries with less stringent climate policies.
4. For the EU ETS, the cap on emissions and price on CO_2 will charge a company for previous free production of by-products and add administrative costs, diverting capital away from other investments.
5. Space does not permit a systematic analysis of lobby activities over time.
6. Non-compliance is not considered a relevant choice, as penalties were set significantly higher than the market price for allowances.
7. The EU introduced a special benchmark regime: free allowances for the most energy-efficient installations.
8. The EU adopted the Market Stability Reserve in 2015, aimed at creating a better balance between supply and demand of allowances and improving resilience to economic fluctuations/shocks. The revised ETS Directive also introduced a gradual 'invalidation' of surplus allowances from 2024,

opportunities for unilateral cancellation of allowances and a more stringent cap of emissions (2.2% annual linear reduction factor as against the current 1.74%).
9. The company argues that Paris is good for gas and business. In 2010, Exxon merged with the US-based XTO Energy, becoming one of the largest gas producers in the world.

Acknowledgments

We are grateful for helpful comments and suggestions from the anonymous reviewers and from Duncan Liefferink and other participants at a workshop on 'Pioneers and Leaders in Polycentric Climate Governance' at the University of Hull in September 2016, organised in the framework of the INOGOV Cost Action Network. This work was supported by the Research Council of Norway (Centre for International Climate and Energy Policy, CICEP) and Nordic Innovation (New Nordic Ways to Green Growth, NOWAGG). The usual disclaimer applies.

Interviews

Chris Beddoes, European Petroleum Industry Association, Europia (15 September 2011, personal).

David Hone, Shell Climate Change Advisor (29 June 2011, phone).

Hans van der Loo, Head European Union Liaison, Shell International (13 April 2011, personal).

Ingvild Skare, Environmental Advisor, ExxonMobil Exploration and Production Norway AS (2 March 2011, phone).

Norbert Herlakian, ExxonMobil, R&S Climate Change Advisor, EMEA Biofuels Venture Mgr. Brussels (12 April 2011, personal).

Jesse Scott, Head of Unit Environment & Sustainable Development Policy, Eurelectric (26 April 2012, personal).

John Scowcroft, Head of Unit Environment and Sustainable Development, Eurelectric (17 November 2010, personal).

Susanne Nies, Head of Unit Energy Policy & Generation, Eurelectric (23 April 2012, personal)

Trym Edvardson, Environmental Discipline Specialist, Shell Upstream International Europe (22 February 2011, phone).

Disclosure statement

No potential conflict of interest was reported by the authors.

References

Ambec, S., et al. 2011. *The Porter hypothesis at 20: can environmental regulation enhance innovation and competitiveness?* Washington, DC: Resources for the Future.
Bach, M., 2019. The oil and gas sector: from climate laggards to climate leaders? *Environmental Politics*, 28, 1.

Bohr, J., 2016. The 'climatism' cartel: why climate change deniers oppose market-based mitigation policy. *Environmental Politics*, 25 (5), 812–830. doi:10.1080/09644016.2016.1156106.

CDP, 2015. *Carbon disclosure project: Exxon's response* [online]. Available from: http://cdn.exxonmobil.com/~/media/global/files/energy-and-environment/2015_cdp_response-pdf.pdf [Accessed 4 June 2018].

Coll, S., 2011. *Private empire: ExxonMobil and American power.* New York: Penguin.

Cyert, R.M. and March, J.G., 1963. *A behavioral theory of the firm.* Englewood Cliffs, NJ: Prentice Hall.

Eikeland, P.O., 2013. Electric power industry. *In*: J.B. Skjærseth and P.O. Eikeland, eds. *Corporate responses to EU emissions trading.* Aldershot: Ashgate, 45–98.

Ellerman, A.C., Convery, F., and de Perthuis, C., 2010. *Pricing carbon: the European Union emissions trading scheme.* Cambridge University Press.

Eurelectric, 2002. *Eurelectric comments to parliament on the commission's proposal for a community greenhouse gas emissions trading scheme.* Brussels: Eurelectric.

Eurelectric, 2007. *Position paper on the review of the EU emissions trading directive (2003/87/EC) and the linking directive (2004/10/EC).* Brussels: Eurelectric.

Eurelectric, 2009a. *A declaration by European electricity sector chief executives* [online]. Brussels: Eurelectric. Available from: http://www.eurelectric.org/CEO/CEODeclaration.asp [Accessed 4 May 2017].

Eurelectric, 2009b. *Power choices—pathways to carbon-neutral electricity in Europe by 2050, full report.* Brussels: Eurelectric.

Eurelectric, 2013. *Consultation on structural options to strengthen the EU emissions trading system: a Eurelectric response.* Brussels: Eurelectric.

Eurelectric, 2016. *Reform of the EU ETS: a Eurelectric statement.* Brussels: Eurelectric.

European Commission, 2006. *EU ETS review: report on international competitiveness.* Brussels: DG Environment.

European Commission, 15 July 2015. *Proposal for a directive of the European parliament and of the council amending directive 2003/87/EC to enhance cost-effective emission reductions and low carbon investments, COM (2015) 337 final.* Brussels: European Commission.

Europia, 2008. *Position paper on the proposed directive amending directive 2003/87/EC.* Brussels: European Petroleum Industry Association.

Europia, 2010a. *White paper on EU refining: a contribution of the refining industry to the EU energy debate.* Brussels: Europia.

Exxon, 2007. *Exxon citizen report.* Available from: http://exxonmobil.com/corporate/ [Accessed 12 June 2018].

ExxonMobil, 2010. *The outlook for energy: a view to 2030* [online]. Available from: http://www.exxonmobil.com/Corporate/files/news_pub_eo.pdf [Accessed 4 June 2018].

Financial Times, 28 March 2017. *Exxon urges Trump to keep US in Paris.*

FuelsEurope, 2014. *Statement on the proposed market stability mechanisms under the EU ETS Directive.* Brussels: FuelsEurope.

FuelsEurope, 2015. *FuelsEurope position on EU ETS Reform.* Brussels: FuelsEurope.

Gravelle, H. and Rees, R., 1981. *Microeconomics.* London: Longman.

Guardian. 2017. The end of coal: EU energy companies pledge no new plants from 2020 [online], 5 April. Available from:: https://www.theguardian.com/environment/2017/apr/05/the-end-of-coal-eu-energy-companies-pledge-no-new-plants-from-2020 [Accessed 3 May 2017].

Handelsblatt Global, 6 October 2016. *Making a renewable giant* [online]. Available from: https://global.handelsblatt.com/companies-markets/making-a-renewable-giant-619138 [Accessed 3 May 2017].

Homsy, G.C. and Warner, M.E., 2015. Cities and sustainability polycentric action and multilevel governance. *Urban Affairs Review*, 51 (1), 46–73. doi:10.1177/1078087414530545.

Kolk, A. and Pinkse, J., 2004. Market strategies for climate change. *European Management Journal*, 22 (3), 304–314. doi:10.1016/j.emj.2004.04.011.

Kolk, A. and Pinkse, J., 2008. A perspective on multinational enterprises and climate change: learning from 'an inconvenient truth'? *Journal of International Business Studies*, 39 (8), 1359–1378. doi:10.1057/jibs.2008.61.

Liefferink, D. and Wurzel, R.K.W., 2017. Environmental leaders and pioneers: agents of change? *Journal of European Public Policy*, 24 (7), 651–668. doi:10.1080/13501763.2016.1161657.

Liefferink, D. and Wurzel, R.K.W., 2018. Leaders and pioneers in polycentric climate governance. *In*: A. Jordan, et al. eds. *Governing climate change: polycentricity in action*. Cambridge: Cambridge University Press.

Meckling, J., 2011. *Carbon coalitions: business, climate politics, and the rise of emissions trading*. Cambridge, MA: MIT Press.

Porter, M. and van der Linde, C., 1995. Toward a new conception of the environment–competitiveness relationship. *Journal of Economic Perspectives*, 9 (4), 97–118. doi:10.1257/jep.9.4.97.

RWE, 2001. *Environmental report 2001*. Essen: RWE.

RWE, 2003. *2003 sustainability report*. Essen: RWE.

RWE, (2013). *Consultation on structural options to strengthen the EU emissions trading system—RWE's response* [online]. Available from: https://ec.europa.eu/clima/sites/clima/files/docs/0017/organisations/rwe_en.pdf [Accessed 4 May 2017].

Simon, H.A., 1976. *Administrative behaviour: a study of decision-making processes in administrative organisation*. 3rd ed. London: Free Press, Collier Macmillan.

Skjærseth, J.B., 2003. *ExxonMobil: tiger or turtle on corporate social responsibility?* FNI Report 7. Lysaker: The Fridtjof Nansen Institute.

Skjærseth, J.B., 2013. Oil industry. *In*: J.B. Skjærseth and P.O. Eikeland, eds.. *Corporate responses to EU emissions trading*. Aldershot: Ashgate, 99–126

Skjærseth, J.B., et al. 2016. *Linking EU climate and energy policies: decision-making, implementation and reform*. Cheltenham: Edward Elgar.

Skjærseth, J.B., 2017. The European Commission's shifting climate leadership. *Global Environmental Politics*, 17 (2), 84–105. doi:10.1162/GLEP_a_00402.

Skjærseth, J.B. and Eikeland, P.O., 2013. *Corporate responses to EU emissions trading*. Aldershot: Ashgate.

Skjærseth, J.B. and Skodvin, T., 2003. *Climate change and the oil industry: common problem, varying strategies*. Manchester: Manchester University Press.

Skjærseth, J.B. and Wettestad, J., 2008. *EU emissions trading: initiation, decision-making and implementation*. Aldershot: Ashgate.

Skjærseth, J.B. and Wettestad, J., 2010. Fixing the EU emissions trading system? Understanding the post-2012 changes. *Global Environmental Politics*, 10 (4), 101–123. doi:10.1162/GLEP_a_00033.

Teece, D.J., 2007. Explicating dynamic capabilities: the nature and microfoundations of (sustainable) enterprise performance. *Strategic Management Journal*, 28 (13), 1319–1350. doi:10.1002/(ISSN)1097-0266.

Tillerson, R., 2010. *Remarks by Tex W. Tillerson, chairman and CEO* [online]. http://www.exxonmobil.com/Corporate/energy_climate_views.aspx [Accessed 12 June 2012].

Vattenfall, 2013. *Vattenfall consultation reply to the European Commission green paper: a 2030 framework for climate and energy policies* [online]. Available from: https://crowdsourcing.simpolproject.eu/static/staticdata/gpc/consultations/vattenfall.pdf [Accessed 4 May 2017].

Vattenfall, 2015. *Vattenfall annual and sustainability report 2014*. Stockholm: Vattenfall.

ⓐ OPEN ACCESS

Cities as leaders in EU multilevel climate governance: embedded upscaling of local experiments in Europe

Kristine Kern ⓘ

ABSTRACT
The success of local climate governance in Europe depends not only on leading cities but also on the dynamics between leaders, followers, and laggards. Upscaling local experiments helps to close the gap between these actors. This process is driven by the increasing embeddedness of cities and their networks in EU multilevel governance. Embedded upscaling combines horizontal upscaling between leading cities with vertical upscaling between leaders and followers that is mediated by higher levels of government, and hierarchical upscaling that even reaches the laggards. Various types of upscaling, their combinations, and their impacts are analyzed. Networks have become denser and networking has intensified. City networks and their member cities have become embedded in national and EU governance, lost authority and depend more and more on regional, national, and European authorities.

Introduction

It is widely acknowledged that cities have become important players in climate governance at national, European, and global levels. At the 2015 Paris climate conference, UN Secretary-General Ban Ki-moon recognized the important role of city leaders, stating that cities have taken leadership to a new level of cooperation and innovation. Many actors from the local to the global level share this view. This development has stimulated many new debates, such as Benjamin Barber's idea to establish a Global Parliament of Mayors (Barber 2013).

Since the Rio Earth Summit, international debates have influenced European cities. Shortly after the Summit in 1992, most leading European cities, such as Copenhagen and Amsterdam, started Local Agenda 21 initiatives, developed indicators, set carbon dioxide (CO_2) reduction targets, and established monitoring systems for measuring their emissions. These pioneering cities not only took early action, but also founded city networks such as Local Governments for Sustainability (ICLEI) and the Climate Alliance to exchange their experiences, competed for awards, branded themselves as green cities, and tried to become models for other cities at home and abroad. Thus, they developed from pioneers, which take action without the ambition to attract followers, into exemplary leaders (Liefferink and Wurzel 2017, Wurzel *et al.* 2019 – this Volume).

Leading European cities not only started earlier than their peers but they have also set more ambitious goals than the EU and its member states. This means that the success of EU energy and climate governance depends not only on the member states but also on subnational action. However, local climate action is not a panacea. Although leading cities have pursued effective climate actions, many cities and towns have not yet introduced appropriate mitigation and adaptation strategies. Despite all the debates on good/best practice transfer and the replicability of experiments, smaller cities and towns may not follow the leaders. On a voluntary basis, good practices are not automatically taken up (Heidrich *et al.* 2016, Reckien *et al.* 2018).

Therefore, the goals of the Paris agreement are attainable only if initiatives are not limited to a few larger cities in metropolitan regions, with the majority of medium-sized and small cities and towns staying behind. Around 40% of Europe's population lives in non-metropolitan regions (Eurostat 2016), and even in metropolitan regions many suburban cities and towns have not developed any relevant strategies. In Germany, for example, only about 30% of the population lives in 80 cities with more than 100,000 inhabitants. Smaller cities and towns have far lower capacities than internationally known leaders. Thus, there is a high potential for CO_2 emission reductions in suburban and rural areas (see also Jänicke and Wurzel – this Volume). As in Germany, the Netherlands, and Sweden, almost all cities with more than 100,000 inhabitants have already started initiatives (for the Netherlands, see den Exter *et al.* 2015); the effectiveness of local climate governance depends on additional actions in smaller cities and towns.

Although even smaller cities (such as Växjö in Sweden) or villages (such as Güssing in Austria) have become internationally known models, such municipalities are 'unlikely pioneers' (Homsy 2018) because the percentage of climate change leaders is highest among large metropolitan regions. Therefore, here I analyse the dynamics between leaders that take action on a voluntary basis and the followers/laggards that require external incentives or even mandatory standards to act (Fuhr *et al.* 2018). Operating on the assumption that system-wide transformation requires climate actions in all municipalities, I ask how cities and towns are governed in a multilevel governance system directly (e.g. by setting mandatory standards) and indirectly (e.g. by certification and rankings), and how cities and towns that have not taken any voluntary action can be stimulated to do so.

I focus on the dynamic relationship between cities in EU multilevel climate governance. Internal and external factors, particularly the embeddedness of local initiatives in polycentric networks of actors at different scales, drive this process. The main argument is that new forms of upscaling have emerged in the EU multilevel governance system. After discussing the nature of my analysis, the research design and the methods, I introduce three types of upscaling of local experiments in multilevel systems: horizontal, vertical and hierarchical upscaling, before suggesting that a new form of upscaling, embedded upscaling, has emerged in the EU multilevel system. I then discuss the impact of embedded upscaling on networking and present some conclusions.

Character of the analysis, research design and methods

This contribution is largely exploratory and conceptual in nature, although it also presents original empirical research. Despite the facts that research on local climate governance has become well-researched, and that interest in large-N analysis (e.g. Reckien, 2018), as well as research on small and medium-sized cities and towns (Hoppe *et al.* 2016, Wurzel *et al.* 2019 – this Volume), have increased considerably, most studies focus on: case studies on leading cities in large metropolitan regions, neglecting mid-sized and smaller cities and towns; global city networks such as C40, neglecting national and European networks and associations of cities; or experiments and urban living labs in larger cities, neglecting the transfer of experiments beyond city borders and to smaller cities and towns in rural areas (Homsy 2018, van der Heijden 2018).

The dynamics between leader, followers, and laggards over time is most evident in relatively affluent liberal democracies that grant local authorities a high degree of political and financial autonomy. I focus on Europe for five main reasons. First, many pioneering and leading cities are located in Europe, and European cities have been at the forefront of taking sustainability and climate policy initiatives. Second, national municipal networks (such as the Dutch *Klimaatverbond)* and transnational municipal networks (such as the Climate Alliance) came into being already 25 years ago. Third, polycentric city networking existed in Europe, in the form of the Hanseatic League, even before the rise of the nation-state. Spurred on by Europe's history of conflicts and wars after the rise of the nation-state, city twinning became popular after the Second World War. Fourth, the first national (subsidy) programs already existed 20 years ago, in particular in the Netherlands and Sweden. In Germany, initiatives started in the federal states (Länder) and, finally, led to an ambitious national program established in 2008. Fifth, the EU Covenant of Mayors (CoM), which the EU Commission initiated in 2008, is a unique feature of multilevel and polycentric governance that goes far beyond transnational city networking. Although the developments in Europe differ from developments in other parts of the world, studying recent trends in Europe may create helpful knowledge for understanding (future) developments outside of Europe.

This contribution builds on original empirical research on climate governance in cities and regions. I conducted around 30 interviews with administrators and politicians in Germany, the Netherlands, and Sweden (in leading cities, regions where these leading cities are located, and national governments), and with representatives of national and international city networks and associations. In addition, I draw on the results of various research projects (Meijering *et al.* 2014, 2018, den Exter *et al.* 2015, Graf *et al.* 2018).

Horizontal, vertical, and hierarchical upscaling in EU multilevel climate governance

Upscaling of local experiments

Although there is widespread interest in upscaling local experiments, no scholarly agreement exists on its definition. Studies on experiments in urban laboratories combine scholarly interest in scales with research on experimental governance (Hoffmann 2011, Castán Broto and Bulkeley 2013, Evans *et al.* 2016). The World Bank (2005), for example, defines upscaling as 'expanding, adapting and sustaining successful policies, programs or projects in different places and over time to reach a greater number of people.'

The concept of scale has been used in various disciplines, ranging from scales in ecosystem management to discussions on 'economies of scale' in economics. The debate on upscaling is most prominent in human geography (van Doren *et al.* 2016, van Winden and van der Buse 2017) and in transition research (Naber *et al.* 2017). Upscaling of local experiments is a process over time that we can characterize by the following:

- *Expansion*: upscaling is limited to the city in which the experiment was conducted, for example, the planned roll-out of a place-based pilot project from one neighborhood to other neighborhoods, driven by project-to-project learning processes;
- *Diffusion*: upscaling between cities on a voluntary basis, based on various forms of networking, ranging from twinning to global city networks;
- *Transformation*: upscaling that leads to a transformation towards sustainability (WBGU 2016) in a specific territory, such as a region or a nation-state, and requires climate action in all municipalities within that territory.

In contrast to existing research on upscaling, which focuses mainly on expansion, i.e. the roll-out of place-based pilot projects (van Doren *et al.* 2016, van Winden and van der Buse 2017), or on socio-technical systems

(Naber *et al.* 2017), here I concentrate primarily on diffusion and transformation in multilevel systems such as the EU and the German federal system. Diffusion of local experiments on a voluntary basis, which does not involve higher levels of government, leads to horizontal upscaling between cities. This facilitates the transfer of good practices to cities and towns that have the capacity to follow the leaders. Transformation towards sustainability in a specific territory requires additional forms of upscaling that involve the state.

I start from the assumption that various types of upscaling exist in multilevel governance systems (Figure 1) (Kern 2014). While *horizontal upscaling* is based on voluntary actions and direct relations between leading cities, *vertical upscaling* is shaped by the interdependent relations between cities and higher levels of government, and *hierarchical upscaling* leads to a harmonization of policies at the national and/or EU level and sets mandatory standards for all municipalities. I claim that a new hybrid mode of upscaling, which I label *embedded upscaling*, is emerging. It combines horizontal, vertical, and hierarchical upscaling (Table 1).

Horizontal upscaling

Horizontal upscaling involves the exchange of experiences, knowledge transfer, and learning between and among cities. Most research on the transfer of good/best practices has focused on debates on policy transfer and diffusion, lesson-drawing, and policy mobility. While policy transfer and lesson-drawing have focused primarily on the transfer of ideas and policies between nation-states, and the discussions on policy diffusion have concentrated on the U.S. states, policy mobility studies primarily analyze

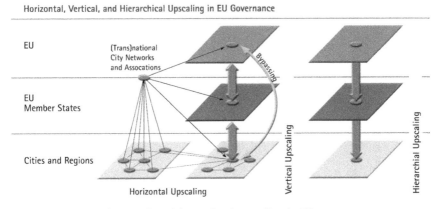

Figure 1. Horizontal, vertical, and hierarchical upscaling in EU governance

Table 1. Types of upscaling.

Type of upscaling	Horizontal upscaling	Vertical upscaling	Hierarchical upscaling	Embedded upscaling
Conceptual approach(es)	(horizontal) policy diffusion; policy transfer, policy mobility, etc.	Multilevel governance	Multilevel governance (Type I)	Multilevel governance (Type II embedded in Type I)
Modes of governance	Governance by diffusion; best-practice transfer; voluntary governance (certification, rankings, and awards); bottom-up approach; crossloading	Governance by (positive) incentives; top-down and bottom-up approaches; uploading	Hierarchical governance; mandatory standards, goals and targets; coercion and sanctions by regional and national authorities; top-down approach; downloading	Combination of horizontal, vertical, and hierarchical modes of governance; polycentric governance
Levels involved	Local level only	EU multilevel system; national multilevel systems (member states)	National multilevel systems (member states)	EU multi-level system; national multilevel systems (member states)
Forms of networking	Twinning; polycentric networking (city networks); main functions: knowledge transfer, exchange of experience	Polycentric networking (city networks, associations of cities and towns); emergence of direct links between the EU and cities (bypassing and scale-jumping); main functions: representation, lobbying, funding	National associations of cities and towns; main functions: representation and lobbying at national level	Polycentric networking: emergence of new forms of networking from regional to EU levels: meta-networks, territorial networks, and functional networks
Leader-follower-laggard dynamics	Learning and transfer among leading cities; widening the gap between leaders and laggards	Regional, national, and EU strategies and programs attract not only leaders but also followers; options for followers to catch up to the leaders	(national) mandates for all cities; even laggards need to comply; closing the gap between leaders and laggards	Combination creates opportunities for leaders, followers, and laggards; closing the gap between leaders and laggards
Challenges	*Experimentation*; experiments are not taken up	*Differentiation*; targeted programs for cities with different ambition levels needed	*Regulation*; Binding standards may have negative repercussions on leading cities	*Integration*; Need to solve the challenges of experimentation, differentiation, and regulation simultaneously

policy mobility between cities (Kern 2000, Karch 2007, Marsden and Stead 2011, McCann and Ward 2012) (see Table 1).

Although scholars and practitioners often refer to the transfer of good practices and the replicability of experiments, there is only limited empirical evidence that place-based experiments actually travel to other places and successfully stimulate policy and institutional changes in other cities at home and abroad. Successful examples include the diffusion of innovations in transport policy, such as Bus Rapid Transport (BRT) initiatives (Marsden and Stead 2011), which started in Latin America and have triggered many initiatives around the world (Mejia-Dugand et al. 2013). In contrast to BRT, highly contested experiments in transport policy, such as congestion charges first introduced in Singapore, show that such experiments may travel and thrive in some places (like London and Stockholm) but are not welcome in others (like New York). Due to many cases of non-diffusion and failed diffusion, adoption rates almost never reach 100%, not even if the diffusion of a policy innovation reaches critical mass and the process becomes self-sustaining (Kern 2000, Kern et al. 2007).

Horizontal upscaling is most relevant for leading cities. Exemplary and cognitive leadership (Wurzel et al. 2019 – this Volume) starts with local experiments that may be replicated within the same city, in other cities in the same country, and in cities in other countries. While experiments are place-based, their transfer depends on polycentric networks that help experiments to cross territorial boundaries and travel to other places.

Leading cities have developed sustainability strategies, integrated climate strategies, and smart city concepts. They have opened their own offices in Brussels and set up city networks, including general networks such as Eurocities, as well as specialized networks such as ICLEI and the Climate Alliance. Leading cities tend to join various networks at the global, European, and national levels, even if these networks fulfill similar functions (interview, City of Freiburg, 2016).

Reckien et al. (2018) found that cities that develop mitigation and adaptation plans are most often large, rich cities with relatively high adaptive capacities that join networks. This group of cities shares certain characteristics. They are: capital cities, such as Paris or Stockholm; second cities, such as Barcelona or Rotterdam; or at least regional centers, such as Hanover, the state capital of the German federal state of Lower-Saxony. These cities are relatively wealthy and powerful with strong research institutions that are highly integrated into the European economy; often, they are close to the sea.

Leading cities are most often located in the Nordic countries (Copenhagen or Stockholm), continental Europe (Amsterdam), and the UK (Bristol). Stockholm became the first European Green Capital in 2010; Copenhagen won this award in 2014, Bristol in 2015, and

Amsterdam was among the finalists in 2010/2011. Leading cities not only join certification systems (such as the European Energy Award), apply for awards, and participate in rankings, they also use their high rankings and their awards to brand the city as a 'sustainable city', 'green city', 'smart city', etc. (Meijering *et al.* 2014, 2018, de Jong *et al.* 2015, Busch 2016). Cities such as Stockholm, Copenhagen, and Freiburg, which are rather small compared to big capital cities like London, do not have the power to act as structural leaders. However, by collecting awards and using their excellent positions in the rankings to brand themselves, they have become intentional exemplary leaders (Liefferink and Wurzel 2018, Wurzel *et al.* 2019 – this Volume), which are acknowledged not only in Europe, but even worldwide.

Only a few cities from southern Europe belong to the leadership group. Exceptions include Vitoria-Gasteiz, which won the European Green Capital Award in 2012, and Barcelona, which was among the finalists in 2012/2013. Cities in central and eastern Europe have shown even less ambition, although this seems to be changing, at least in countries that have started to set up national programs, such as the Polish low-carbon economy plan (Donnerer 2016). Due to challenges after the fall of the Berlin Wall, only a few cities in central and eastern Europe have joined transnational networks, rarely competing for awards, and, in rankings, the cities at the very bottom are most often located east of the former Iron Curtain (Siemens 2009). There are a few exceptions, such as the pioneering Polish city of Bielsko-Biala or Ljubljana, the European Green Capital of 2016.

Horizontal upscaling among leading cities gains support from the following dynamics: bilateral city twinning, i.e. long-term networking of a rather general nature that can provide a basis for more complex forms of cooperation; project networking of a limited number of cities, which facilitate tailor-made forms of knowledge transfer and learning; and multilateral networking of cities, particularly (trans)national city networks. Leading cities joined at least one of three transnational city networks, i.e. the Climate Alliance, Energy Cities, and ICLEI, which pioneering cities founded in the early 1990s. From the outset, the exchange of experiences, transfer of knowledge, and stimulation of learning among their members crystalized as one of their key functions (Kern and Bulkeley 2009, Fünfgeld 2015, Busch 2016). Membership in these networks grew rapidly in the first years but slowed when these networks matured and became more consolidated. Today, it has become difficult to attract new member cities in Europe.[1] In contrast, the development of global city networks, such as C40, seems to be far more dynamic (for discussion on global city networks see Bouteligier 2013, Bansard *et al.* 2016, Gordon and Johnson 2017, 2018), but only 18 European cities are members of the C40 network. Most of these are capital cities, and not all are leading cities (for example Rome, Moscow, and Istanbul) (cf. van der Heijden 2018).

Since horizontal upscaling is most prominent among leading European cities, it is not surprising that researchers and practitioners alike have focused on leading cities and horizontal upscaling. As we can characterize most European cities as followers and laggards, horizontal upscaling is a necessary first step, but it is not sufficient for system-wide transformations because potential followers may not have the capacities required to follow the leaders, but can only do so if they get external support provided by governmental and non-governmental actors, e.g. through national funding programs or the establishment of new agencies that provide services and advice (see also Wurzel et al. 2019 – this Volume).

Vertical upscaling

Upscaling of local experiments is not limited to horizontal upscaling between leading cities because the role of cities in EU climate governance has changed. Authority and competencies shifted not only upwards to the EU, but also downwards to subnational authorities (see Hooghe and Marks 2003, Monni and Raes 2008, Emilianoff 2014, Jänicke and Quitzow 2017, Jänicke and Wurzel 2019, – this Volume). Initiatives range from the development of new institutions, such as local and regional energy agencies, to guidelines and new funding programs. In several member states, national strategies and guidelines guide local climate policy (Heidrich et al. 2016), while other member states have developed subsidy programs. Funding projects and offering choices between various ambition levels enable smaller cities and towns, with less capacity and lower ambitions than the leading cities, to start climate actions (see Table 1).

If there is a lack of appropriate national programs, cities may turn their attention to EU programs. EU funding programs are most welcome, even by leading cities such as Amsterdam or Malmö (interviews: City of Hanover, City of Freiburg, City of Amsterdam, all 2016; Stumpp 2016). Therefore, cities have developed new strategies to get access to EU institutions, for example by bypassing national authorities (see Figure 1). Going to Brussels generates new opportunities for cities. These strategies are in line with research on Europeanization that has shown that leading countries influence EU decision-making and try to upload their policies to the European level, so they become binding for all member states, including the laggards (Börzel 2002).

I characterize the relationship between the EU and cities as involving interdependent relations and polycentric networking. Vertical upscaling requires that city networks and associations represent their members and lobby at regional, national, and EU levels. Apart from a few big cities with structural power and leadership (Liefferink und Wurzel 2018) that have the means to represent their interests directly, the strategies of city networks and associations become decisive (Monni and Raes 2008, Kern 2014). Thus, the Climate Alliance, Energy Cities, and ICLEI have developed active strategies to lobby for the interest of their member cities in Brussels. There are various venues from which to influence EU institutions, including the Committee of the Regions (CoR).

As the EU Commission has an interest in cooperating with cities in a more systematic way, it supports their activities in Brussels. An early example is the Commission's support of the European Sustainable Cities and Towns Campaign. This initiative started in 1994 and attracted, in particular, Spanish and Italian cities (Echebarria *et al.* 2004, Sancassiani 2005). Today, the Campaign consists of five transnational networks and associations of local authorities, such as Eurocities and the Council of European Municipalities and Regions (CEMR); a committee of representatives of cities, the EU Commission, and the EU Expert Group on the Urban Environment coordinate the campaign.

Taking not only leading metropolitan cities but also smaller cities and towns into account requires a stronger focus on national and regional associations that represent all cities and towns in a given territory. The Council of European Municipalities and Regions (CEMR) represents smaller cities and towns in Brussels. At the national level, regional and national networks of cities and towns (such as the Swedish *Klimatkommunerna* or the Dutch *Klimaatverbond*) may be far more important than transnational city networks (see Table 1).

Since vertical upscaling is not limited to leading cities, but facilitates the transfer of innovations between leaders and followers, the analysis of vertical upscaling requires better understanding of municipalities that are neither leaders nor laggards, and the dynamics between the leaders and this group of cities and towns. Vertical upscaling provides incentives for cities and towns that are not (yet) at the forefront of local climate action but want to start such initiatives and catch up with the leaders. However, in the absence of hard regulations, there are still a considerable number of municipalities that are not taking any action on a voluntary basis.

Hierarchical upscaling

I characterize hierarchical upscaling as initiatives at European, national, and regional levels, which force the laggards to reach standards set by the EU and its member states. In contrast to horizontal and vertical upscaling, hierarchical upscaling requires strong governments with the authority and power to harmonize policies and set binding standards. Relations between different levels of government are organized top–down, and authority concentrates at EU and member state levels (see Figure 1 and Table 1), while the authority of cities is limited to the implementation of EU and national legislation.

Decisions made in Brussels or in the national capitals of the member states affect all local authorities in the EU. Traditionally, environmental policy incorporated standard-setting and a command-and-control style of policy-making that left implementation to subnational authorities. Hierarchical governance plays a decisive role in the development of EU environmental policy, for example, by setting binding emission standards for air pollutants, but this has often resulted in implementation deficits.

Hierarchical governance is far less developed in EU climate governance than in other policy areas. Local climate policy is still a voluntary task in most EU member states. National and regional governments face a challenge here because hard mandates are not always an option, due to the fact that local authorities have the right to local self-government, on the one hand, and no means to comply with such mandates, on the other.

As voluntary actions by leading cities and their followers most often do not reach the laggards, I argue that horizontal and vertical upscaling need to be complemented by hierarchical upscaling, i.e. binding rules for all municipalities. Despite all actions of leading cities and soft policies, I expect that laggards become active only if mandatory requirements exist. From an upscaling perspective that takes the dynamics between leaders, followers, and laggards over time into account, hierarchical upscaling is a process that starts with local experiments in leading cities. Their ideas and experiences are taken up by the national government (vertical diffusion), transformed into national regulations, and, finally, become binding for all municipalities.

In the EU, binding regulations for municipalities are still limited to a few member-states. Local climate plans are required only in France, the UK, Slovakia, and Denmark. In France, the central government requires intermunicipal authorities with more than 20,000 inhabitants to develop local climate and energy plans (Donnerer 2016), and, in the UK, the Climate Change Act demands that local authorities integrate climate mitigation and adaptation policies in their local planning documents (Heidrich *et al.* 2016, Reckien *et al.* 2018). In Scotland, the 2009 Climate Change Scotland Act even sets general GHG emission reduction targets for Scottish cities.

While climate mitigation depends mainly on voluntary actions at the local level, binding standards can be found more often in energy policy, for instance, energy efficiency standards for (new) buildings. Based on EU directives, particularly the Energy Performance of Buildings Directive that requires that new buildings deliver nearly zero-energy consumption by 2020, all member states must enact mandatory standards for new buildings.

In Germany, for example, the Energy Savings Law *(EnergieeinsparungsG)* and the Energy Savings Ordinance *(Energieeinsparverordnung, ENEV)* accomplished this. After a revision in 2016, the ENEV even states that almost all buildings should be climate-neutral by 2050. Leading cities, such as Freiburg and Heidelberg, developed specific competences within local government and initiated their own programs (interview, Climate Protection and Energy Agency Baden-Württemberg, 2016) long before binding standards were set. These initiatives facilitated the enactment of national and state regulations for the energy efficiency of buildings. However, the implementation of these regulations in all municipalities has led to serious implementation deficits due to a lack of capacity in many smaller municipalities (Graf et al. 2018).

Moreover, hierarchical upscaling may restrict the leaders if national standards are binding and do not allow leading cities to set stricter standards. This situation is avoidable only by setting minimum standards, i.e. standards that are binding (for the laggards) but nonetheless allow leading cities to set higher standards on a voluntary basis (Table 1).

Embedded upscaling

The challenges of horizontal, vertical, and hierarchical upscaling caused the emergence of embedded upscaling as a new hybrid form of upscaling (Table 1). Embedded upscaling means that Type II multilevel governance (Hooghe and Marks 2003), characterized by task-specific and intersecting membership and a flexible design, is embedded in Type I multilevel governance, i.e. in general-purpose, multi-functional, non-intersecting jurisdictions (EU, national governments, regions) (Liefferink and Wurzel 2018). Moreover, embedded upscaling also shows the main elements of polycentric governance, which 'seems to be the key concept in addressing the complexity of territorial planning and management in Europe' (Finka and Kluvánková 2015, p. 606). This is in line with Elinor Ostrom's argument that polycentric systems with multiple governing authorities at different scales have advantages due to mechanisms for learning, adaptation, and mutual monitoring (Ostrom 2010, p. 552). Embedded upscaling links a variety of governing authorities at different scales, offers new options for experimentation and learning, not restricted to leaders, and polycentric networking becomes embedded in existing governance systems.

Both the EU Covenant of Mayors (CoM) and the German *Kommunalrichtlinie (KRL)* are forms of embedded upscaling. The KRL program, established in 2008, has funded around 12,500 projects in more than 3,000 German municipalities (around 25% of all German municipalities). Funding is obtainable, for example, for investments in energy-efficient street lighting, climate protection concepts, and climate management (interview, German Institute for Urbanism, 2016). It is particularly interesting for poor municipalities because they can get higher subsidies. In addition, the German federal government has also supported 41 leading municipalities *(Masterplankommunen, MPK)*. Two cohorts of MPKs, selected on a competitive basis in 2012 and 2016, received funding for four years and have committed themselves to reduce GHG emission by 95% by 2050.

Dynamic interactions between the federal government, state governments, and (leading) cities have accompanied the development of the KRL. Today, this program's implementation involves close cooperation by the Federal Environment Ministry; the Service and Competence Center for Local Climate Protection *(Service- und Kompetenzentrum: Kommunaler Klimaschutz)*, which is affiliated with the German Institute for Urban Affairs *(Deutsches Institut für Urbanistik)*; and Project Management Jülich *(Projektträger Jülich)*.

At EU level, the EU Commission (DG Energy), supported by the Committee of the Regions and the EU Parliament, set up the *Covenant of Mayors (CoM)* at almost the same time. Its main aim has been the local implementation of the EU Climate and Energy Package of 2008. Thus, signatories committed themselves to reducing their CO_2 emissions by at least 20% by 2020. In March 2014, the EU Commission complemented the CoM with Mayors Adapt, a second initiative that the EU Commission (DG Climate Action) launched in cooperation with the European Environment Agency. In the fall of 2015, both initiatives merged and became the *Covenant of Mayors for Climate & Energy*. Signatories are obliged to develop integrated strategies to tackle climate mitigation and adaptation and reduce their CO_2 emissions in line with the EU's 40% target by 2030. 7,755 local authorities with almost 253 million inhabitants had joined the initiative by August 2018, among them many small and medium-sized cities and towns in Italy and Spain. Networks and associations of local and regional authorities run the CoM Office, the Intelligent Energy Europe program provides funding, and the EU Commission's Joint Research Center assesses all action plans and monitoring reports (Kona *et al.* 2015). Signatories have already submitted around 6,000 action plans and around 1,700 monitoring reports.

The increasing embeddedness of initiatives in multilevel governance systems is most obvious with respect to the CoM because the Covenant differs considerably from traditional city networks. It is a unique institutional arrangement based on the close cooperation of all major European city networks, the EU Commission, and the European Commission's Joint Research Centre as the monitoring agency. The CoM gets support from more than 200 *Covenant Coordinators* (national and regional authorities such as Italian provinces) and more than 180 *Covenant Supporters* (national and regional city networks and associations, local and regional energy agencies).

The percentage of cities that participate in the CoM differs considerably between member states. While around 40% of all Italian municipalities have joined the CoM, in Germany or France, less than one percent of the municipalities have signed. In countries with a high number of small municipalities, these differences decrease when attention is shifted from the number of participating municipalities to the population covered by the CoM: 23% of Germans, 26% of French, and 70% of Italians live in municipalities that joined the CoM. The differences between Germany and Italy are explicable by the parallel development of the EU CoM and the German KRL; both initiatives started in 2008 and fulfill similar functions because they facilitate and stimulate climate action in mid-sized cities and towns. Due to the KRL and the initiatives funded by the federal government, German cities and towns do not see added value in joining the CoM. Participation in the CoM seems to be limited to the leading cities, and even these pioneers complain about the extra burden of monitoring CO_2 emissions in different ways because no harmonized system of monitoring exists (interviews: City of Hannover and City of Freiburg, 2016, City of Potsdam, 2017, Donnerer 2016). The missing linkages between the CoM and the KRL also contribute to low participation of German cities in the CoM. Such linkages exist in countries with a high number of CoM signatories; for example, the provinces of Valencia (Spain) and Vlaams-Brabant (Belgium) allocate additional financial support for municipalities that join the CoM (Donnerer 2016).

Upscaling and networking

Embedded upscaling has impacts on city networks, which were established by leading cities, for leading cities (Kern and Bulkeley 2009). Based on the examples of the KRL and the CoM, I argue that embedded upscaling changes the characteristics of city networks. In the 1990s, European city networks attracted primarily cities that considered themselves to be leaders or wanted to become leaders, while the CoM attracts not only many small cities and towns in Italy and Spain but also municipalities that want to act but do not intend to become vanguard cities. The examples of the CoM and the KRL show that embedded upscaling is based on various forms of cooperation and networking, including new practices, new actors, and new networks.

First, the CoM developed into a *meta-network*, i.e. a network which has other networks and associations as members. At least in Europe, this is a general trend. Energy Cities, for example, has around 200 individual member cities and 20 collective members (such as the Dutch *Klimaatverbond* and the Union of the Baltic Cities) with around 2,600 member cities and towns. The CoM is supported by almost 100 associations and networks of local and regional authorities (such as the Association of Polish Cities and the Climate Alliance Austria). In addition, the CoM office in Brussels is funded by the EU and run by a consortium of all major, and sometimes competing, networks and associations of local and regional authorities in Europe.[2] The emergence of meta-networks has advantages for the EU Commission because all relevant networks need to speak with one voice, but it also means that networks are forced to cooperate with each other and become more dependent on the EU.

Second, at national and subnational levels, we find at least three types of *territorial networking*:

- networking at the national level, initiated and driven by local authorities such as *Klimatkommunerna* in Sweden, and national/regional associations that represent all cities and towns in a given territory (interviews: Rijkswaterstaat, 2017; Klimatkommunerna, 2017);
- networking initiated and driven by regional authorities (e.g. Italian provinces or German counties) that make initiatives such as the CoM or the KRL work on the ground, for example, by coordinating and supporting the development of Joint Sustainable Energy Actions Plans (SEAPs) for smaller municipalities (Rivas *et al.* 2015, De Gregorio Hurtado *et al.* 2015, interviews Climate and Energy Agencies Lower Saxony and Baden-Württemberg, 2016);
- networking initiated and driven by actors at the national and EU levels, such as the National Clubs of the CoM, i.e. national networks of CoM signatories.

Third, new *functional networks* have emerged as a byproduct of funding programs: for example, networks of climate protection managers in Germany. The German federal government indirectly created such positions at local level through KRL funding. In Baden-Württemberg, the federal state funds the establishment of regional energy agencies (most of them at county level) that now cover nearly the whole state and have led to a dense network of cooperation among these new institutions (interview, Climate and Energy Agency Baden-Württemberg, 2016). The federal government and the federal states (Länder) initiate and support networking between these new actors (interview, German Institute for Urbanism, 2016). This has resulted in new forms of functional networking because these young and engaged employees are a rather homogenous group and fulfill similar functions in their municipalities. In 2016, climate protection managers even founded their own professional association *(Bundesverband Klimaschutz)*.

To sum up, embedded upscaling is accompanied by the emergence of various new types of networking at different scales, ranging from the EU to the regions. This includes meta-networking, which has become an important task of European city networks. Additionally, new forms of territorial networking at the national and regional levels, and functional networking among climate managers or regional energy agencies have emerged.

Conclusions

Local climate action has become an important feature of European climate governance, and a considerable percentage of Europeans now live in cities with relatively ambitious reduction goals. Although big, wealthy, and powerful cities, led by charismatic leaders have become important players in climate governance, local climate action is not a panacea. While many cities have reduced their CO_2 emissions considerably, most small and medium-sized municipalities in Europe have not yet started taking climate initiatives.

Most leading European cities are located in the Nordic countries, continental Europe, and the UK. Only a few leaders can be found in southern and eastern Europe. Research has focused on leading cities (Copenhagen, Amsterdam, and Freiburg, for example), horizontal upscaling, and global city networks. Much less is known about smaller cities and towns; researchers have neglected new forms of upscaling and networking that are more important for followers and laggards. The success of local climate policy in Europe depends, however, not only on a small group of leading cities, but also on follower cities that are willing to catch up with the leaders, and on binding standards for laggards that would otherwise not start their own initiatives.

Cities have become embedded in European multilevel governance. Polycentric governance by embedded upscaling goes far beyond the voluntary upscaling of local experiments, and combines certain elements of horizontal, vertical, and hierarchical upscaling. Embedded upscaling helps to bridge the gap between leaders, followers, and laggards and provides tools for differentiated approaches that are needed because municipalities differ considerably. While leading cities have a European or even international orientation, followers and laggards are nationally or even regionally oriented. As leading cities are rare in southern and eastern Europe, specific strategies for these regions are needed. While the Covenant of Mayors seems to work well in southern Europe, where regional authorities support cities and towns, there is still a lack of local climate initiatives and regional support in the former socialist countries.

The EU Covenant of Mayors and the German *Kommunalrichtlinie* are examples of embedded upscaling by polycentric governance. Both initiatives coordinate and orchestrate multiple governing authorities at different scales. Rule-based, incorporating monitoring systems, and allowing for experimentation and learning, embedded upscaling provides options for leading cities such as Copenhagen to maintain their leadership role and distinguish themselves from the rest of the pack, for followers to catch up with the leaders or at least improve their performance at a lower ambition level, and for laggards if they want to catch up.

Embedded upscaling creates new opportunities for cities, the EU, and its member states because it establishes new forms of networking. First, *meta-networking* has emerged at EU level. Although climate leadership by cities has always been supported by city networks, these networks have changed their form and function. The CoM has generated incentives for follower cities, but has also had repercussions on city networks that have become more dependent on EU institutions. Second, both case studies show the importance of *territorial networking*. Upscaling place-based experiments is most successful if they become connected to regional and national networks, that are more important for mid-sized cities and towns (see also Wurzel et al. 2019 – this Volume). In southern Europe, the success of the CoM depends on the integration of regional authorities in the CoM, and in Germany the involvement of counties and regional energy agencies are decisive for the successful implementation of the KRL. Third, the German KRL shows an increase of *functional networking* because it led to the emergence of networks of local climate managers at regional and national scale.

This increase of meta-networking, territorial networking, and functional networking shows that (polycentric) networks have become denser and more intense. The inclusion of followers and laggards may cause changes for city networks and their members. The cases of the CoM and the KRL show that cities and their networks have become embedded in EU and national climate governance. City networks that were developed bottom–up have lost authority and depend more and more on regional, national, and European authorities.

Notes

1. While Energy Cities and the Climate Alliance have grown during the last 15 years, ICLEI lost members in Europe. Since 2000 ICLEI membership in Germany has not only decreased by 25% but also shows a high degree of fluctuation. Around 60% of the cities that were members of this network in 2000 have left. The Climate Alliance has grown, but most new members are small municipalities in Austria. As almost 90% of its members are German and Austrian cities, today's Climate Alliance is far less international than it was in its early days. However, this also means that 37% of all Austrian municipalities are members of the Climate Alliance, which cooperates closely with all Austrian federal states *(Bundesländer)* that joined the Climate Alliance as associated members.

2. Energy Cities, Climate Alliance, Council of European Municipalities and Regions (CEMR), Eurocities, European Federation of Agencies and Regions for Energy and the Environment (FEDARENE), and ICLEI Europe.

Acknowledgments

This work draws mainly on results of the research project *Post-Carbon Cities of Tomorrow (POCACITO)*, which was funded by the EU under the 7th Framework Programme.

Disclosure statement

No potential conflict of interest was reported by the author.

Funding

This work was supported by the EU 7th Framework Programme [grant number 613286].

ORCID

Kristine Kern http://orcid.org/0000-0001-9923-4621

References

Bansard, J., Pattberg, P., and Widerberg, O., 2016. Cities to the rescue? Assessing the performance of transnational municipal networks in global climate governance. *International Environmental Agreements*, 17 (2), 229–246. doi:10.1007/s10784-016-9318-9
Barber, B., 2013. *If mayors ruled the world: dysfunctional nations, rising cities*. New Haven & London: Yale University Press.
Börzel, T., 2002. Member state responses to Europeanization. *Journal of Common Market Studies*, 30 (2), 193–214. doi:10.1111/1468-5965.00351
Bouteligier, S., 2013. *Cities, networks, and global environmental governance*. New York & London: Routledge.
Busch, H. 2016. *Entangled cities. Transnational municipal climate networks and urban governance*, doctoral dissertation, Lund University.
Castán Broto, V. and Bulkeley, H., 2013. A survey of urban climate change experiments in 100 cities. *Global Environmental Change*, 23 (1), 92–102. doi:10.1016/j.gloenvcha.2012.07.005
De Gregorio Hurtado, S., et al., 2015. Understanding how and why cities engage with climate policy: an analysis of local climate action in Spain and Italy. *TEMA: Journal of Land Use, Mobility and Environment*, 8, 23–46.
de Jong, M., et al., 2015. Sustainable – smart – resilient – low carbon – eco – knowledge cities: making sense of a multitude of concepts promoting sustainable urbanization. *Journal of Cleaner Production*, 109, 25–38. doi:10.1016/j.jclepro.2015.02.004

den Exter, R., Lenhart, J., and Kern, K., 2015. Governing climate change in Dutch cities: anchoring local climate governance in organization, policy and practical implementation. *Local Environment*, 20 (9), 1062–1080. doi:10.1080/13549839.2014.892919

Donnerer, D. 2016. *The role of cities in climate and energy policies: the case of the Covenant of Mayors*, Master thesis, International Studies, Aarhus University.

Echebarria, C., Barrutia, J.M., and Aguado, I., 2004. Local Agenda 21: progress in Spain. *European Urban and Regional Studies*, 11 (3), 273–281. doi:10.1177/0969776404041490

Emilianoff, C., 2014. Local energy transition and multilevel climate governance: the contrasted experiences of two pioneer cities (Hanover, Germany, and Växjö, Sweden). *Urban Studies*, 51 (7), 1378–1393. doi:10.1177/0042098013500087

Evans, J., Karvonen, A., and Raven, R., eds., 2016. *The experimental city*. London & New York: Routledge.

Eurostat, 2016. *Urban Europe. Statistics on cities, towns and suburbs*, 2016 edition. Luxembourg: Publications office of the European Union. https://ec.europa.eu/eurostat/documents/3217494/7596823/KS-01-16-691-EN-N.pdf/0abf140c-ccc7-4a7f-b236-682effcde10f

Finka, M. and Kluvánková, T., 2015. Managing complexity of urban systems: a polycentric approach. *Land Use Policy*, 42, 602–608. doi:10.1016/j.landusepol.2014.09.016

Fuhr, H., Hickmann, T., and Kern, K., 2018. The role of cities in multi-level climate governance: local climate policies and the 1.5 C target. *Current Opinion in Environmental Sustainability*, 30, 1–6. doi:10.1016/j.cosust.2017.10.006

Fünfgeld, H., 2015. Facilitating local climate change adaptation through transnational municipal networks. *Current Opinion in Environmental Sustainability*, 12, 67–73. doi:10.1016/j.cosust.2014.10.011

Gordon, D. and Johnson, C., 2017. The orchestration of global urban climate governance: conducting power in the post-Paris climate regime. *Environmental Politics*, 26 (4), 694–714. doi:10.1080/09644016.2017.1320829

Gordon, D. and Johnson, C., 2018. City-networks, global climate governance, and the road to 1.5°C. *Current Opinion in Environmental Sustainability*, 30, 35–41. doi:10.1016/j.cosust.2018.02.011

Graf, P., Kern, K., and Scheiner, S., 2018. Mehrebenen-Dynamiken in der deutschen Energiewendepolitik. Die Rolle von Städten und Regionen am Beispiel von Baden-Württemberg. *In*: K. Radke and N. Kersting, eds. *Energiewende. Politikwissenschaftliche Perspektiven*. Wiesbaden: Springer, 211–247.

Heidrich, O., et al. 2016. National climate policies across Europe and their impact on city strategies. *Journal of Environmental Management*, 168, 36–45. doi:10.1016/j.jenvman.2015.11.043

Hoffmann, M.J., 2011. *Climate governance at the crossroads: experimenting with a global response*. Oxford: Oxford University Press.

Homsy, G., 2018. Unlikely pioneers: creative climate change policymaking in smaller U.S. cities. *Journal of Environmental Studies and Sciences*, 8 (2), 121–131. doi:10.1007/s13412-018-0483-8

Hooghe, L. and Marks, G., 2003. Unraveling the central state? Types of multi-level governance. *American Political Science Review*, 97 (2), 233–243.

Hoppe, T., van der Vegt, A., and Stegmaier, P., 2016. Presenting a framework to analyze local climate policy and action in small and medium-sized cities. *Sustainability*, 8 (9), 847. doi:10.3390/su8090847

Jänicke, M. and Quitzow, R., 2017. Multi-level reinforcement in European climate and energy governance: mobilizing economic interest at the sub-national levels. *Environmental Policy and Governance*, 27 (2), 122–136. doi:10.1002/eet.1748

Jänicke, M. and Wurzel, R., 2019. Leadership and lesson-drawing in the European Union's climate governance system. *Environmental Politics*, 28 (1).

Karch, A., 2007. *Democratic laboratories. Policy diffusion among the American states*. Ann Arbor: University of Michigan Press.

Kern, K., 2000. *Die Diffusion von Politikinnovationen. Umweltpolitische Innovationen im Mehrebenensystem der USA*. Opladen: Leske + Budrich.

Kern, K., 2014. Climate governance in the EU multi-level system: the role of cities. *In*: I. Weibust and J. Meadowcroft, eds. *Multilevel environmental governance: managing water and climate change in Europe and North America*. Cheltenham: Edward Elgar, 111–130.

Kern, K. and Bulkeley, H., 2009. Cities, Europeanization and multi-level governance: governing climate change through transnational municipal networks. *Journal of Common Market Studies*, 47 (1), 309–332. doi:10.1111/j.1468-5965.2009.00806.x

Kern, K., Koll, C., and Schophaus, M., 2007. The diffusion of Local Agenda 21 in Germany: comparing the German federal states. *Environmental Politics*, 16 (4), 604–624. doi:10.1080/09644010701419139

Kona, A., et al., 2015. *The covenant of mayors in figures and performance indicators: 6-year assessment*, European Commission: Joint Research Centre. https://www.conventiondesmaires.eu/IMG/pdf/Covenant_of_Mayors_in_Figures_and_Performance_Indicators_6-year_Assessme—.pdf

Liefferink, D. and Wurzel, R., 2018. Leadership and pioneership. Exploring their role in polycentric governance. *In*: A. Jordan, D. Huitema, H. van Asselt, and J. Foster, eds. *Governing climate change: polycentricity in action?* Cambridge: Cambridge University Press, 135–151.

Liefferink, D. and Wurzel, R., 2017. Environmental leaders and pioneers: agents of change?. *Journal of European Public Policy*, 24 (7), 951–968. doi:10.1080/13501763.2016.1161657

Marsden, G. and Stead, D., 2011. Policy transfer and learning in the field of transport: a review of concepts and evidence. *Transport Policy*, 18 (3), 492–500. doi:10.1016/j.tranpol.2010.10.007

McCann, E. and Ward, K., 2012. Assembling urbanism: following policies and 'studying through' the sites and situations of policy making. *Environment and Planning A*, 44 (1), 42–51. doi:10.1068/a44178

Meijering, J., Kern, K., and Tobi, H., 2014. Identifying the methodological characteristics of European green city rankings. *Ecological Indicators*, 43, 132–142. doi:10.1016/j.ecolind.2014.02.026

Meijering, J., Tobi, H., and Kern, K., 2018. Defining and measuring urban sustainability in Europe: A Delphi study on identifying its most relevant components. *Ecological Indicators*, 90, 38–46. doi:10.1016/j.ecolind.2018.02.055

Mejía-Dugand, S., et al., 2013. Lessons from the spread of bus rapid transit in Latin America. *Journal of Cleaner Production*, 50, 82–90. doi:10.1016/j.jclepro.2012.11.028

Monni, S. and Raes, F., 2008. Multilevel climate policy: the case of the European Union, Finland and Helsinki. *Environmental Science and Policy*, 11 (8), 743–755. doi:10.1016/j.envsci.2008.08.001

Naber, R., et al., 2017. Scaling up sustainable energy innovations. *Energy Policy*, 110, 342–354. doi:10.1016/j.enpol.2017.07.056

Ostrom, E., 2010. Polycentric systems for coping with collective action and global environmental change. *Global Environmental Change*, 20 (4), 550–557. doi:10.1016/j.gloenvcha.2010.07.004

Reckien, D., et al., 2018. How are cities planning to respond to climate change? Assessment of local climate plans from 885 cities in the EU-28. *Journal of Cleaner Production*, 191, 207–219. doi:10.1016/j.jclepro.2018.03.220

Rivas, S., et al., 2015. *The Covenant of Mayors: in-depth analysis of sustainable energy action plans*. European Commission: Joint Research Centre. https://www.pattodeisindaci.eu/IMG/pdf/2015-11-13_JRC_SEAPAnalysis.pdf

Sancassiani, W., 2005. Local Agenda 21 in Italy: an effective governance tool for facilitating local communities' participation and promoting capacity building for sustainability. *Local Environment*, 10 (2), 189–200. doi:10.1080/1354983052000330770

Siemens, 2009. *European green city index. Assessing the environmental impact of Europe's green cities*. Munich: Siemens AG. https://www.siemens.com/entry/cc/features/greencityindex_international/all/en/pdf/report_en.pdf

Stumpp, I. 2016. *The role of national government initiatives in urban climate governance: the case of Malmö*, Master thesis, Potsdam: University of Potsdam.

van der Heijden, J., 2018. City and subnational governance: high ambitions, innovative instruments and polycentric collaborations? *In*: A. Jordan, D. Huitema, H. van Asselt, and J. Foster, eds. *Governing climate change: polycentricity in action?* Cambridge: Cambridge University Press, 81–96.

Van Doren, D., et al., 2016. Scaling-up low-carbon urban initiatives: towards a better understanding. *Urban Studies*, 55 (1), 175–194. doi:10.1177/0042098016640456

Van Winden, W. and van der Buse, D., 2017. Smart city pilot projects: explaining the dimensions and conditions of scaling up. *Journal of Urban Technology*, 24 (4), 51–72. doi:10.1080/10630732.2017.1348884

WBGU, 2016. *Humanity on the move: the transformative power of cities*. Berlin: German Advisory Council on Global Change, Berlin.

World Bank. 2005. Reducing poverty, sustaining growth: scaling up poverty reduction. Case study summaries, Conference in Shanghai, May 25-27, 2004.

Wurzel, R., et al., 2019. Climate pioneership and leadership in structurally disadvantaged maritime port cities. *Environmental Politics*, 28 (1).

Wurzel, R., Liefferink, D., and Torney, D., 2019. Pioneers, leaders and followers in multilevel and polycentric climate governance. *Environmental Politics*, 28 (1).

Climate pioneership and leadership in structurally disadvantaged maritime port cities

Rüdiger K.W. Wurzel ⓘ, Jeremy F.G. Moulton, Winfried Osthorst, Linda Mederake, Pauline Deutz and Andrew E.G. Jonas

ABSTRACT
Innovative climate governance in small-to-medium-sized structurally disadvantaged cities (SDCs) are assessed. Considering their deeply ingrained severe economic and social problems it would be reasonable to assume that SDCs act primarily as climate laggards or at best as followers. However, novel empirical findings show that SDCs are capable of acting as climate pioneers. Different types and styles of climate leadership and pioneership and how they operate within multi-level and polycentric governance structures are identified and assessed. SDCs seem relatively readily willing to adopt transformational climate pioneership styles to create 'green' jobs, for example, in the offshore wind energy sector and with the aim of improving their poor external image. However, in order to sustain transformational climate pioneership they often have to rely on support from 'higher' levels of governance. For SDCs, there is a tension between learning from each other's best practice and fierce economic competition in climate innovation.

Introduction

International relations (IR) and comparative politics (CP) scholars initially dominated the research on climate leaders and pioneers, focusing mainly on climate governance at the international, supranational and state level (e.g. Gupta and Grubb 2000). At first, scholars paid little attention to climate leaders and pioneers at sub-state level (e.g. regional, local and city levels) although there are exceptions, especially in the local governance literature (e.g. Bulkeley and Betsill 2005, While *et al.* 2010). Over time, the importance of local climate governance increased to such a degree that Jänicke (2014, p. 43) has argued that 'the local level is a late mover in the process of climate policy, but has become the most dynamic driver of technical change towards a low-carbon energy system'.

The importance of local climate governance has increased for at least three reasons. First, IR and CP scholars discovered the significance of cities for global climate governance when international climate negotiations threatened to end in political stalemate. Secondly, cities are both major sources of greenhouse gas emissions (GHGE) and laboratories for experimentation with innovative learning-by-doing climate governance measures, some of which could possibly be up-scaled from the local level to 'higher' governance levels (e.g. Bulkeley *et al.* 2015, Kemmerzell 2017, Eckersley 2018, Kern 2019, – this volume); thirdly, there is the issue of state 'hollowing out' where states have lost power *upwards* (to the international and/or supranational level), *sideways* (to business and societal actors) and *downwards* (to the subnational level) (e.g. Strange 1998). Social scientists have tried to capture the purported move from top-down climate *government* towards bottom-up climate *governance* analytically with the help of multilevel governance (MLG) concepts (e.g. Schreurs and Tiberghien 2007, Wurzel *et al.* 2017) and polycentric governance approaches (e.g. Ostrom 2009, 2014, Morrison *et al.* 2017, Jordan *et al.* 2018, Singleton 2018) in which non-state and subnational actors play a prominent role. Urban politics scholars especially have argued that environmental governance is being rescaled around local and regional state structures in response to wider political and economic restructurings in liberal market economies (While *et al.* 2010). However, the nation-state level continues to play an important role for many climate governance initiatives by structurally disadvantaged cities (SDCs), which are not unitary actors. As we explain below, local level societal climate alliances play an important role for the ability of SDCs to adopt innovative climate governance measures

Although there is growing interest in local climate governance, much of the literature has focused on relatively affluent and/or large cities that have acted as climate leaders or pioneers (e.g. Jonas *et al.* 2011, Bulkeley *et al.* 2015) and their national and transnational city networks (e.g. Kern and Bulkeley 2009). It has paid little attention to innovative local climate governance by less affluent small-to-medium-sized cities such as SDCs which suffer from serious economic resource constraints (for exceptions see Bulkeley *et al.* 2015, Jonas *et al.* 2017, Eckersley 2018).

Here, we focus on Bremerhaven (Germany) and Hull[1] (UK) as case study cities because both classify as SDCs faced with similar economic, social and geographic challenges. Moreover, Bremerhaven and Hull are maritime port cities that have perceived climate change as both a threat (flooding due to sea water level rise) and an opportunity (offshore wind energy jobs). Both Bremerhaven and, to a lesser degree, Hull have experimented with innovative local climate governance measures. They have done so while acting primarily as what Liefferink and Wurzel (2017) define as pioneers. Liefferink and Wurzel (2017, see also Wurzel et al. 2017) have argued that leaders actively try to attract followers while this is not normally the case for pioneers. Considering their acute resource constraints and other structural disadvantages, it seems reasonable to assume that SDCs act primarily as climate laggards or at best as followers. However, based on our novel empirical findings, we argue that this is not necessarily the case. We show that SDCs are capable of acting as climate pioneers and, to a lesser degree, leaders. This creates a research puzzle, which we try to explain by answering the following main research question: How and why do SDCs act as climate pioneers or leaders?

We proceed as follows. In the next section, we define SDCs before briefly reviewing the urban climate governance literature while linking it to the analytical concept of leaders and pioneers put forward by Liefferink and Wurzel (2017). We then assess Bremerhaven and Hull's main innovative local climate governance activities. The penultimate section uses MLG and polycentric governance concepts to analyse the empirical findings from our two cities. The concluding section reassesses our conceptual framework and the main empirical findings while offering general conclusions about small-to-medium-sized SDCs.

Small-to-medium-sized structurally disadvantaged cities

For our definition of SDCs, we draw on Jonas et al. (2017) who have defined structurally disadvantaged maritime port cities as suffering from: geographical remoteness; long-term decline of industries (e.g. for maritime port cities fishing and shipbuilding); disused industrial assets and infrastructure (e.g. port facilities); high unemployment, low/underutilised skills base and declining populations; weak economic governance structures (including shrinking tax bases and high susceptibility to austerity measures); and, poor external image. Both Bremerhaven and Hull share these characteristics.

Our concept of SDCs resembles theories of structurally disadvantaged communities that have assessed minority communities in American inner cities, which exhibit serious social pathologies such as crime and public disorder (Kane 2005). Building on Wilson's (1987) seminal book, *The Truly Disadvantaged*, our concept seeks to capture the social and institutional structures (race and class) that have resulted in such communities becoming economically and socially marginalised. By way of contrast, we use the term SDCs in a more spatially and socially encompassing fashion to refer to small-to-medium-sized cities that are grappling with a range of structural problems. In other words, our focus is not on a particular social subgroup *within* the city but rather a collective of small-to-medium-sized cities that exhibit industrial decline, population loss, social problems, geographical peripherality and negative external images.

There is no hard and fast definition for what constitutes a medium-sized city. Le Galés (2002, p.5) has defined medium-sized cities as having a population of between 150,000 and 200,000 inhabitants. However, Le Galés (2002, p.32) also cites Kaelble (1988, p.62) who has pointed out that about one-third of Europe's population has lived in '[t]he medium-sized town of between 20,000 and 100,000 inhabitants [which has] played a more significant and more enduring role in the twentieth century than elsewhere'. Small-to-medium-sized cities cover free-standing urban centres having a population between 100,000 and 500,000; these are autonomous or separate local political jurisdictions governed by an elected city council or magistrate and led by a locally appointed or elected mayor; usually they are second or third tier urban centres within their host national economy/state. We have adopted the term small-to-medium-sized city to take into account that Bremerhaven had a population of about 112,000 while Hull consisted of approximately 258,000 inhabitants in 2016. The city of Bremerhaven together with the city of Bremen forms the state *(Land)* Bremen while Hull belongs to the Humber region; they have populations of about 672,000 and 921,000, respectively.

For many years, Bremerhaven has been among those German municipalities suffering the most severe economic problems (Wegweiser Kommune 2016). Röhl and Schröder (2017) even concluded that Bremerhaven is the poorest German city in terms of purchasing power. In 2005, Bremerhaven suffered from over 25% unemployment; in 2016, it was still 14.6%, more than twice the German national average (Statistisches Landesamt Bremen 2017). Bremerhaven eventually reversed decades of population decline in the early 1990s. In 1968, Bremerhaven's population peaked at about 149,000 before bottoming out at about 108,000 in 2011. By 2016, Bremerhaven's population had grown again to about 116,000, largely due to international immigration (Statistisches Landesamt Bremen 2017). Although Bremerhaven qualifies as a SDC, it nevertheless functions as a regional center for the labour market. In 2015, commuters constituted 47.3% of regular employees in Bremerhaven (Bertelsmann Stiftung 2017). Commuters who work in Bremerhaven but are residents in Lower Saxony *(Land* Niedersachsen), which surrounds the cities of Bremerhaven and Bremen, do not pay taxes in the *Land* Bremen.

Hull's population declined for decades after peaking at around 302,000 in 1931, falling to below 244,000 in 2001. The 2011 census found the decline had moderately reversed, largely due to international immigration (Migration Observatory 2014). However, compared to 2016 8% fewer migrants arrived in Hull in 2017 (Migration Yorkshire 2018). This decline was largely due to the outcome of the UK's 2016 Brexit referendum. Hull's citizenry suffers from inter-generational unemployment, lack of skills development and social exclusion. In 2016, the city's unemployment rate was around 7.4% compared to 4.6% UK-wide, which constituted a significant improvement compared to 13.5% (7% national average) in 2015 (Hull Data Observatory 2017)[2]. A 2014 study of 64 UK cities, which compared indicators such as earnings, job seekers allowance and employment, ranked Hull amongst the most problematic cities (Centre for Cities 2014).

Cities and climate pioneership and leadership

'Cities lie at the heart of the challenge of addressing climate change' (Bulkeley et al. 2015, p.5) because they produce large amounts of GHGE and can act as laboratories for experimentation with innovative climate change measures. In the 1990s, a limited number of relatively large and/or prosperous cities (acting as leaders or pioneers that formed national and/or transnational networks) largely drove innovative urban climate governance (Kern and Bulkeley 2009, Kern 2019, – this volume). Largely ignoring SDCs, early urban climate governance studies focused mainly on leading cities and their best practices, core indicators and success factors, while emphasising the increasing relevance of local climate governance for urban 'green' economic development.

Liefferink and Wurzel (2017) have argued that while leaders usually actively seek to attract followers, this is not normally the case for pioneers. They furthermore distinguished between four *types* of leadership/pioneership – structural, entrepreneurial, cognitive, and exemplary – and two *styles* of leadership/pioneership – transactional and transformational (Wurzel

Table 1. Core features of Bremerhaven and Hull.

	Bremerhaven	Hull
Inhabitants (2016)	c. 116,000 (*Land* Bremen c. 667,000)	c. 258,000 (Humber region c. 921,000)
Unemployment (2015)	15.1% (national average 6.4%)	13.5% (national average 7%)
Industrial decline	Fishing industry, ship building, departure of US army	Fishing industry, food industry
Current major employers	Offshore wind energy industry, port, logistics, food industry, research (e.g. AWI), Bremerhaven University	Chemical industry, port, University of Hull, offshore wind energy industry
Offshore wind energy industry	Adven/Areva and Senvion/RePower as well as WeserWind (until its insolvency) and PowerBlades (until it relocated)	Siemens, A2Sea
Research related to offshore wind energy	Fraunhofer Institute for Wind Energy, Alfred-Wegener-Institut, Hochschule Bremerhaven	University of Hull (e.g. aura project), Hull College Energy and Climate Centre
Direct jobs in offshore wind energy industry	c. 3,500 in 2014; c. 1,500 in 2017 (Germany: c. 20,000 in 2017)	c. 1,000 in 2017 (c. 10,000 in UK in 2017)
Major flooding	1962, 1999, 2006, 2016	1953, 2007, 2013

Sources: Jonas et al. (2017), Hull Data Observatory (2017*)* and Statistisches Landesamt Bremen (2017).

et al. 2017). However, their leadership/pioneership concept focused only on states while largely ignoring the subnational level. We apply Liefferink and Wurzel's (2017) analytical environmental leader and pioneer concepts to cities while focusing on SDCs.

In IR and CP, *structural* leadership/pioneership is associated primarily with military and economic power. While military power does not play a significant role for tackling climate change, economic power capabilities are crucial at any governance level including the local. In contrast to small-to-medium-sized cities (e.g. Bremerhaven and Hull), large cities (e.g. Berlin and London) have considerable economic power and thus structural leadership/pioneership capabilities (see Kern 2019, Jänicke and Wurzel 2019, – both this volume). However, as Burns (1978, p. 19) has pointed out, '[a]ll leaders are actual or potential power holders, but not all power holders are leaders'. Cities' formal institutional powers can be interpreted as representing structural power resources. For example, city states *(Stadtstaaten)* such as Bremen, which encompasses the cities of Bremerhaven and Bremen, have significant powers under the German federal constitution. Structural leadership/pioneership activities may also include economic actions aimed at improving cities' positions in urban hierarchies or vis-à-vis cities with similar 'green' economy ambitions.

Entrepreneurial leadership/pioneership involves the use of diplomatic and negotiating skills with a view to brokering integrative bargains and agreements. Entrepreneurial leadership also includes networking between actors, sectors and governance levels. As explained below, for an emerging new industrial sector like the offshore wind industry, entrepreneurial leadership in the form of networking is of central importance. MLG concepts have tried to capture analytically such networks and their interdependencies across different governance levels. While emphasising the importance of local entrepreneurial initiatives, experimentation and learning-by-doing, polycentric climate governance concepts (Ostrom 2009, 2014) have stressed the significance of self-organisation, trust building and site-specific conditions (e.g. Morrison *et al.* 2017, Jordan *et al.* 2018).

Cognitive leadership/pioneership involves defining and redefining interests and developing innovative ideas such as the 'green economy' or low carbon economy, which aim to generate 'green' jobs while reducing GHGE. Cognitive leadership/pioneership may conceptualise climate change not only as a threat but also as an opportunity. It may also relate to branding/rebranding strategies with the aim of improving the external image of SDCs (e.g. from climate laggard to leader or pioneer), which try to attract inward investment and skilled people to the city. Such rebranding goes well beyond superficial 'greenwashing' or symbolic climate leadership/pioneership if significant GHGE reductions support it. The urban governance literature has emphasised the importance of cities and regions as laboratories for experimentation and innovation (e.g. Bulkeley and Betsill 2005, Ostrom 2009) although little is known about whether SDCs can fulfill similar functions.

Finally, *exemplary* leadership/pioneership refers to the setting of examples for others, either intentionally or unintentionally. *Intentional* exemplary leaders put forward climate governance measures as models for others. *Unintentional* example setting, in contrast, refers to pioneers who do not intentionally aim to attract followers (Wurzel *et al.* 2017). Cities that adopt the Covenant of Majors usually try to set a good example that they would like others to follow (e.g. Bulkeley *et al.* 2015). Such cities therefore act as climate leaders. However, adoption of innovative urban climate governance measures may involve no intention of setting an example for other cities, thus amounting to climate pioneership rather than leadership.

Liefferink and Wurzel (2017) differentiate between *internal* and *external* ambitions, arguing that an actor with high internal and low external ambitions acts as a *pioneer* with no explicit intention to attract followers (see also Wurzel *et al.* 2017 – this volume). We argue that SDCs may well have high internal climate ambitions that they do not normally use to attract followers. We therefore use the term pioneership to refer to internal climate ambitions of SDCs (rather than their external ambitions which would amount to leadership).

These four different *types* of leadership/pioneership can be combined analytically with the following two leadership/pioneership *styles*, namely *transactional* and *transformational* (Liefferink and Wurzel 2017). Transactional leadership/pioneership refers to incremental changes usually over a relatively short time horizon while transformational leadership/pioneership aims at profound or even 'revolutionary' changes usually over a comparatively long time period. Transactional climate leadership/pioneership resembles efforts to make cities more resilient to climate change although such efforts can lead to a 'resilience trap' (Kythreotis and Bristow 2017) that merely reinforces the status quo. However, transactional leadership/pioneership extending over a very long timescale may eventually also trigger transformational change (e.g. Burns 1978).

Climate governance in structurally disadvantaged cities

Urban climate strategy

In Bremerhaven, local climate governance experienced a substantial boost and institutionalisation with the approval of the Climate City (*Klimastadt*) concept in 2010. Its adoption followed a motion in the city parliament (*Bremerhavener Stadtverordnetenversammlung*) in 2007 and the publication of a conceptual study in 2009 (AWI 2009). The full name of the *Klimastadt* is *Kurs Klimastadt*, which translates as 'on course to becoming a climate city' and 'resonates well with the maritime image of Bremerhaven' (Interview, 2016). As one climate city office *(Klimastadtbüro)* staffer emphasised: 'We are not yet a climate city. We are on course to becoming one' (Interview, 2016). As we explain below, Bremerhaven's climate city concept identified not only municipal steering and monitoring instruments but also proposed the creation of an innovative green economy cluster that, until the mid-2010s, focused almost exclusively on the offshore wind energy industry.

In 2006, Bremerhaven followed Bremen's lead by applying for the European Energy Award (EEA) certification scheme to establish a monitoring and implementation tool for climate management. Five years later, Bremerhaven achieved the scheme's requirements and obtained the EEA award. In 2007, Bremerhaven prepared a research and development concept with the aim of transforming the existing network of climate-related institutions into a flagship project while trying to exploit its full economic potential. The concept focused on the following three 'climate lighthouses' *(Klimaleuchttürme)*: regional business promotion in the offshore wind energy sector; top-level, climate-related research activities; and tourist attractions such as the climate house *(Klimahaus)* museum (Mederake 2015). Initiated by a Christian Democratic Union (CDU) and Social Democratic Party (SPD) local government coalition in 2008, Bremerhaven City Council adopted a Master Plan Active Climate Policy as strategic frame for local state and non-state climate measures.

In 2010, Bremerhaven launched the climate city Bremerhaven *(Klimastadt Bremerhaven)* initiative with the aim of improving its poor external image, strengthening its climate-related capacities, boosting jobs in the offshore wind energy sector and reducing GHGE. At the centre of the climate city project was the attempt to raise Bremerhaven's internal climate ambitions with the aim of changing its negative image in order to attract external investment. Bremerhaven adopted important infrastructure measures while providing advice to potential investors through the Bremerhaven Economic Development Company and City Development *(Bremerhavener Gesellschaft für Investititionsförderung und Stadtentwicklung- BIS)*, which increased its staff and knowledge resources on offshore wind energy. Bremerhaven thus positioned itself as a climate pioneer while offering especially entrepreneurial and cognitive pioneership. With the adoption of its climate city project, Bremerhaven opted for a transformative pioneership style, which was aimed at mitigating climate change while transforming the city's economic fortunes with the help of the offshore wind energy industry.

Following local elections for Bremerhaven's City Council in 2011, a 'Red-Green' coalition government of the SPD and the Greens replaced the CDU-SPD coalition. The 2011 local elections had occurred shortly after the

Fukushima nuclear catastrophe, which greatly boosted the electoral support for the Greens who secured (among other posts) the Environmental Councillor *(Umweltdezernentin)*. At the request of the Greens, the focus of the *Klimastadt* project widened from a relatively narrow focus on 'green' business activities (the offshore wind energy sector) to include also public participation initiatives with civil society actors. Largely on the insistence of the new Green Environmental Councillor, the climate city office *(Klimastadtbüro)*, which had been set up in 2014, moved to new premises in a prime location in the city centre where it opened its doors for the general public in 2015. However, the need for budget cuts and changed political priorities of the CDU-SPD government newly elected in Bremerhaven in 2015 triggered a moderate reduction in staff and the *Klimastadtbüro's* relocation to cheaper, more remote premises in 2017.

Bremerhaven's local climate governance targets became more ambitious as a consequence of programmes developed by the *Land* Bremen, which a SPD-Greens coalition has governed since 2007. The *Land* Bremen adopted the Climate Protection and Energy Policy Programme 2020 *(Klimaschutz- und Energieprogramm 2020* – KEP 2020) for Bremen and Bremerhaven in 2010 (SUBV 2010). The KEP set an ambitious 40% reduction target for CO_2 by 2020 (compared to 1990). Consequently, Bremerhaven also committed itself to reducing CO_2 emissions by 40% by 2020, reduction targets that were strengthened when the Bremen State Climate and Energy Act 2015 included them *(Bremisches Klimaschutz- und Energiegesetz)*.

Hull's Environment & Climate Change Strategy 2010–2020, published in 2010, also set ambitious CO_2 emissions reduction goals. It noted the EU's legally binding 20% and the UK's 32% reduction goals for 2020 and adopted the somewhat more ambitious goal of between 32% and 45% emissions reductions by 2020 (Hull City Council 2010). Since the publication of the 2010 Strategy, funding cuts imposed by central government on UK local government have had an adverse effect on the willingness and ability of Councils to fund climate action (Eckersley 2018) and their ability to offer climate pioneership. However, funding cuts have promoted independent power generation from renewable sources and energy efficiency measures to save money.

Since 2011, Hull Council has published annual CO_2 emissions reports. Efforts to reduce emissions include the deployment of photovoltaic panels on Council buildings and the replacement of street lighting with LED bulbs. Hull City Council partnered with 'green economy' and third sector actors to create the Green City Group in 2011 to consider the branding of Hull as a *Green City*, in a similar but coincidental manner to Bremerhaven's *Klimastadt* project. Ultimately, with the adoption of Hull's City Plan the Council made the decision to market Hull instead as *Energy City* (Hull City Council 2013). Whilst this represented an ideational shift away from outright climate pioneership to an economic marketing strategy, the City Council has nevertheless demonstrated some local climate pioneership. Unlike in Bremerhaven, one political party (Labour) has dominated post-Second World War local politics in Hull, with the exception of 2007-11 when the Liberal Democrat Party held the majority. Other political parties, such as the Green Party and the United Kingdom Independence Party (UKIP), have had little influence in local government.

Especially when considering the vulnerability of maritime port cities to climate change (e.g. sea water level rise) and SDCs' resource constraints, it is perhaps surprising that Bremerhaven and Hull adopted a significant number of climate mitigation initiatives instead of solely supporting adaptation activities. Post-Second World War Bremerhaven has suffered from significant flooding events in 1962, 1999, 2006 and 2016, while Hull has been affected by major flooding in 1953, 2007 and 2013 (see Table 2). Recent serious flooding events have helped local officials in both cities to push climate change higher up the local government agenda (Interviews, 2014-2017). As one Hull City Council official stated, flooding has been key in developing the city's climate actorness: 'From the experience in 2007 with the floods there and the 2013 tidal surge, we're a lot further ahead than a lot of cities'.

Societal participation strategies
Bremerhaven's *Klimastadt* programme set up six working groups in which local government, civil society, and business actors cooperate on: economy/science, citizens and education, construction and modernisation of buildings, communication, mobility, and sustainable tourism. Additional innovative participatory elements in the *Klimastadt* project included an annual festival-like climate day *(Klimatag)* and, since 2014, a Youth Climate Council *(Jugendklimarat)* which has a small budget and the right to speak in the City's Environmental and Construction Committee. In 2013, the *Klimastadt* funded the transdisciplinary festival *Odyssee Klima* (Odyssey climate) in which Bremerhaven's city theatre *(Stadttheater)* took on a leading role. The festival featured climate change related plays and performances by actors and scientists in Bremerhaven's city centre and in wind turbine production factories (Interviews, 2014-17).

Hull's status as the UK's City of Culture in 2017 involved few climate-related activities. When compared to Bremerhaven, societal participation has remained underdeveloped in Hull's climate governance strategy. Hull's adoption of the *Energy City* brand aimed mainly at business and 'green economy' actors rather than at societal actors. However, Hull used its status as the UK's City of Culture in 2017, albeit hesitantly, to raise public attention to climate change-related issues (Interview, 2016). One example, 'The Blade' installation, involved exhibiting a 75-metre rotor blade of an

Table 2. Climate governance innovations in Bremerhaven and Hull.

	Bremerhaven	Hull
Urban climate strategy		
Strategic manage-ment	• 2008 MAK, updated 2011 • 2010 KEP: −40% CO_2 by 2020 • Bremen State Law on Climate Protection and Energy: −40% CO_2 by 2020 • EEA since 2006 • Environment Protection Agency: one official for climate and the Climate City Office since 2014	• Environment and climate change strategy 2010–2012: −34% CO_2 by 2020; ambition: −45% • Green City Group (2011–2012) • Councillor with Energy City portfolio • One Council official responsible for environ-ment/climate
Societal participation	• Six Climate City working groups • Youth Climate Council • Climate City Day since 2013 • Urban climate policy par-ticipatory dialogue (2012) • Climate change education at schools	• Goodwin Development Trust's climate-themed artist residencies and community engagement projects
Rebranding	• Climate City concept since 2009 • Odyssee Climate event (2013) • Electric cars rally (2016)	• Energy City concept • Humber region as 'Energy Estuary' • 2017 City of Culture cli-mate-themed events
Green economy		

Offshore Wind Energy	• Offshore wind energy cluster: peak of c.3,500 jobs in 2014; declined to 1,500 in 2017
	• Research institutes (e.g. Alfred-Wegener-Institut, Fraunhofer Institute for Wind Energy Systems)
	• Training and research at Hochschule Bremerhaven
	   ◦ WAB
	   ◦ BIS
	   ◦ OTB plans stalled
		• Green Port Hull: c. 2,000 jobs in 2017; moderate increase expected
		• Siemens c. 1,000 jobs in 2017
		• Specialised training (e.g. Humber Onshore & Offshore Training Association)
		• Training and research at Hull University (e.g. pro-ject aura)

offshore wind turbine in the city centre in early 2017. Another was an exhibition entitled 'Somewhere becoming Sea' which focused on the ever-changing boundaries between land and sea while trying to capture 'the sea's elemental power …[a]t a time when climate change threatens to blur boundaries further and bring far-reaching economic impact' (Hull UK City of Culture 2017).

The green economy

In Bremerhaven, the *Land* Bremen and the city of Bremerhaven identified the offshore wind energy industry and, to a much lesser degree, the onshore wind energy industry as potential major growth sectors in 2003. This occurred against the background of ambitious national renewable energy targets, generous subsidies for renewable energy and the decision to phase out nuclear power in Germany. There was a concerted attempt to turn easy access to the sea, disused industrial facilities (e.g. from shipyards) and derelict land together with underutilised maritime-related job skills into an advantage. Substantial investment and various state and city agencies (including the BIS) facilitated the creation of an offshore wind energy cluster in Bremerhaven. Highly regarded, nationally funded research facilities such as the Alfred-Wegener-Institut (AWI) and Fraunhofer Institute for Wind Energy and Energy System Technology (IWES) moved to Bremerhaven in 1980 and 2009, respectively. Sectoral associations and specialised networks for the offshore wind energy sector (e.g. the wind energy agency Bremerhaven (*Windenergie Agentur Bremerhaven* – WAB)) also came into being. The WAB became an important network, which provided entrepreneurial pioneership in the form of networking opportunities for the fledgling offshore wind energy industry in Bremerhaven and the wider region. The University of Applied Sciences Bremerhaven (*Hochschule Bremerhaven*) offered specialised academic training programmes, thus strengthening Bremerhaven's cognitive pioneership capacities. In short, the local government in Bremerhaven tried to create, with the support of in particular businesses, entrepreneurial and cognitive pioneership capacities for the offshore wind energy industry. Its structural pioneership capacities relied primarily on economic support from the *Land* Bremen and/or federal government funds.

In 2014, the offshore wind energy industry reached a peak with around 3,500 jobs in Bremerhaven (Written communication, BIS, 2017). The joint efforts of Bremerhaven and the *Land* Bremen led to the establishment of a leading offshore renewable energy industry cluster that included companies such as Adven (formerly Areva) and Senvion (previously RePower) as well as WeserWind, which became insolvent in 2015, and PowerBlades which decided to closed its factory in Bremerhaven in 2018. Arguably the concerted action by Bremerhaven in cooperation with the *Land* Bremen amounted to local government structural leadership with the aim of transforming the city's socio-economic structure.

As a result of the offshore wind energy industry boom, the Senate of the Land Bremen adopted plans for an Offshore Terminal Bremerhaven (OTB) in Bremerhaven at an estimated cost of €180 million in 2010. However, they were halted at least temporarily when the environmental NGO, BUND (Friends of the Earth Germany), took legal action against its construction, arguing that the OTB was no longer economically viable. As a consequence of reforms to the German Renewable Energy Law *(Erneuerbare-Energien-Gesetz* – EEG) in 2014 and 2016, funding for offshore wind energy production decreased significantly. The federal government also adopted a two-year moratorium for the expansion of offshore wind farms in the North Sea while government funding was diverted for political reasons to offshore wind farms in the Baltic Sea, which economically benefit the northern coastal state in the former East Germany. The combination of these measures significantly dampened hopes for ambitious expansion plans for Bremerhaven's offshore wind energy capacity. The number of direct and indirect jobs in the offshore wind energy sector in Bremerhaven, which peaked at approximately 3,500 in early 2014, fell to about 1,500 staff in 2017 (Written communication, BIS, 2017). The main reason for this steep decline was that WeserWind, which had employed approximately 1,200 staff in 2012–2013, went bankrupt in early 2015. In the same year Siemens, one of Europe's leading offshore wind turbine producers, invested in a new factory in neighbouring Cuxhaven while its wind turbine production facilities across the North Sea in Hull became operational in early 2017. Moreover, Powerblades, which had a staff of about 300 in Bremerhaven in 2017, announced plans to relocate its plant in early 2018. Consequently, Bremerhaven has tried to broaden its relatively narrow focus on the offshore wind energy industry towards a wider focus on the green economy (Interviews, 2017).

Hull's climate pioneership is mainly the result of the adoption of green economy measures and a municipal drive towards climate change mitigation and adaptation. Compared to Bremerhaven, Hull's adoption of climate governance measures is more hesitant. Whilst Hull exhibited climate pioneership somewhat later than Bremerhaven, its green economy measures have been less narrowly focused on the offshore wind energy sector while including also other forms of renewable energy such as biomass. Hull has become a centre for the biofuel industry, which is however not without its environmental critics (e.g. Mol 2010). Associated British Ports (ABP) along with the Spencer Group, a Hull-based engineering company, developed a biomass terminal and storage facility for fuel for the formerly coal-fired Drax power plant (in North Yorkshire). Within Hull itself, Vivergo Biofuels, a co-venture by BP, AP Sugar, and DuPont, was the largest biorefinery in the UK when it opened.

Despite the decline of the fishing industry, a section of Hull's port-based industry stayed active. The ports act both as cargo and ferry terminals. However, the gap left by the decline of the city's primary industry led to the recognition that a new major industry was needed to give the city 'a renewed *raison d'etre*' (Interview, 2016). It is into this gap that Hull actors ushered the green economy. Here 'gap' refers to both an economic and physical space which made Hull an attractive site for Siemens when considering the placement of a new blade manufacturing and offshore wind turbine assembly facility. As well as the brownfield and quay space in Hull, the city's location close to important UK offshore wind farm sites, made it a particularly suitable development location (Jonas *et al.* 2017). Hull City Council, local MPs and national politicians were able to secure the Siemens facility as part of a £310 million joint venture with ABP which manages four important regional ports: Hull, Goole, Grimsby, and Immingham. The Siemens-ABP facility opened in early 2017. From initial plans to produce 450 blades a year, Siemens has increased its intended yearly output to 600 blades for deployment in 6MW turbines in UK waters and further afield. However, the UK's decision to leave the EU has put in question Siemens' future ability to use its Hull production site for exports to the EU. By 2017 the site created 1,000 direct jobs, 95% of whom were from the region. The Siemens-ABP Green Port Hull development is Hull's most high-profile green economy development, but there are also other local green economy projects.

Spurred on by the achievement of attracting Siemens to Hull, local economic development practitioners have seen an opportunity to address the city and region's longstanding structural disadvantages. In other words, local governance actors have tried to adopt a transformative pioneership style. In 2013, Hull approved a 10-year City Plan to attract £1bn in investment and create up to 8,000 new jobs for local job seekers over the next 10 years (Hull City Plan, 2017). A central component of the City Plan was *Energy City*, an umbrella term for flagship industry projects – the majority of which can be classed as part of the green economy. A key driver in Hull's proposed economic transformation has been the Humber Local Enterprise Partnership (LEP). In 2013, LEPs replaced Regional Development Agencies as private sector-led economic development organisations, albeit retaining significant public sector representation. In its *Strategic Economic Plan*, the Humber LEP recognised that addressing climate change is essential to the competitiveness of Hull because of the need for building inward investor confidence and reducing flood risk. Nevertheless, although the LEP has recognised the wider social and environmental benefits of climate adaption, its priority is to support mainstream economic development, including 'green' jobs. LEP's main goal is to put 'the Humber [at] the centre of renewable energy [in the UK] – so that when people think of energy they will think of the Humber' (Interview, Hull, 2014).

The evolution of the renewables sector in Hull demonstrates the city's belated structural pioneership, which could turn out to have been an advantage. The city has become home to an established, market-dominant player – Siemens. If Hull had pushed for such an offshore wind production base earlier, it possibly would have been harder to find such an established partner. The insolvency of WeserWind and the relocation of Powerblades in Bremerhaven as well as Siemens' investment in neighbouring Cuxhaven illustrate the risk that structural local climate pioneership poses for SDCs. The 'first mover' advantage, which has been identified for states and companies (e.g. Porter 1990), may possibly play out differently at the city level – at least for SDCs. Hull closely followed Bremerhaven's exemplary pioneership. By sending delegations to Bremerhaven, Hull tried to learn lessons from one of Europe's offshore wind energy pioneers. However, lesson-drawing quickly turned into competition for investment – as Bremerhaven was also a city considered for the Siemens plant now housed in Hull. This shows that there is a thin dividing line between learning from best practice generated by local climate experimentation and fierce economic competition between SDCs wanting to attract investment from transnational corporations (Kemmerzell 2017).

Explaining urban climate governance in SDCs: multi-level, polycentric or place-specific?

MLG concepts have emphasised the mutual dependency of governance actors at different governance levels (e.g. EU and subnational levels). In contrast, polycentric governance concepts usually attribute a higher degree of autonomy to subnational actors (e.g. cities and regions) and societal actors (e.g. business, NGOs and individual citizens) as regards experimentation with innovative local climate governance measures and learning-by-doing. Put simply, while MLG concepts tend to focus on the globalization (or 'glocalisation') and, within the European political context, the Europeanisation of regional and local actors (e.g. Hooghe 1996), polycentric governance concepts stress the crucial role that local climate governance experiments play for the success of global climate governance regimes (e.g. Ostrom 2009).

Polycentric governance concepts share certain core presuppositions (e.g. multiple centres of authority and levels of governance) with MLG approaches, although conceptually they are not identical (Homsy and Warner 2015, Wurzel et al. 2017, Jordan et al. 2018). By comparison with MLG approaches, polycentric concepts normally assume a stronger role for societal actors and attribute a high degree of autonomy to both subnational actors (e.g. cities) and non-governmental societal actors (Ostrom 2009, 2014). Polycentric governance approaches usually emphasise the importance of bottom-up local governance and argue that local climate governance ought to supplement, if not partly supplant, global climate governance initiatives. Broadly speaking, proponents of polycentricity favour societal self-coordination within market-like governance structures (e.g. Ostrom 2014, for critical reviews see Morrison et al. 2017, Singleton 2018) while MLG advocates support the creation of networks in which governmental actors (including supranational EU actors) play an important, if not dominant, role, e.g. to correct negative market externalities (Hooghe 1996, Homsy and Warner 2015).

Proponents of polycentricity usually argue in favour of a multitude of decision-making 'centres' and widespread subnational societal self-coordination in climate governance (e.g. Ostrom 2009, 2014). From a polycentric perspective, one would normally expect smaller and more independent cities to exhibit greater degrees of climate leadership/pioneership. At first sight, the empirical findings presented above might suggest that it is 'game, set and match' for polycentric governance perspectives as Bremerhaven, which has exhibited a higher degree of climate innovation, is smaller and enjoys a greater degree of local governance independence compared to Hull. However, our empirical findings also show Bremerhaven's high dependency (especially for structural climate pioneership) on decisions that were taken on the national governance level. Prominent examples are the reforms of the EEG and its detrimental impact on the offshore wind energy industry in Bremerhaven. Our findings therefore emphasise the importance of MLG structures and concepts.

Both Bremerhaven and Hull took account of national and EU CO_2 reduction targets when setting their local reduction targets. As explained above, for Bremerhaven the CO_2 reduction targets of the *Land* Bremen were also of crucial importance while for Hull no such additional layer of local climate governance existed.

Conclusion

Here, we have assessed how small-to-medium-sized SDCs (Bremerhaven and Hull) have responded to climate change by developing climate pioneership capacities at the urban level. Based on our empirical evidence, Bremerhaven exhibited a higher degree of experimentation and innovation in local climate governance than Hull, which focused more on the mainstreaming of innovative local climate action within economic development, with the latter remaining dominant. Bremerhaven's climate city *(Klimastadt)* concept clearly goes well beyond Hull's Energy City concept in terms of local climate governance innovation. While offering exemplary pioneership, Bremerhaven has adopted slightly more ambitious medium-term CO_2 reduction goals than Hull .

Hull initially acted as a follower in relation to Bremerhaven in terms of creating offshore wind energy industry capacities although it overtook Bremerhaven in the late 2010s. There are concerns about Bremerhaven's ability to maintain climate pioneership over a long time period, which is usually required to produce transformative effects, without additional support from a 'higher' governance level (e.g. the

Land Bremen, federal government and/or EU). The long-term success of Bremerhaven's transformative pioneership may also require some modification, which is not unusual for learning-by-doing local climate experiments and innovation. For example, Bremerhaven's initial narrow focus on the offshore wind energy sector as a transformative industrial sector had to be broadened to encompass the wider green economy following significant changes at the national governance level (e.g. the reforms of the EEG) which had a detrimental effect on the offshore wind energy industry in Bremerhaven.

As we have focused only on two case study cities, we are able to draw only tentative general conclusions about innovative climate governance in SDCs, which have remained under-researched. Further research is necessary to show whether our empirical findings are indeed representative for SDCs. Such research ought also to assess critically whether the use of the pioneer and leader concept indeed adds analytical value to critically assessing local climate governance. With this in mind, we put forward the following five main conclusions.

First, considering their structural disadvantages it would be reasonable to assume that SDCs act primarily as climate laggards or at best as followers. However, our empirical research has clearly shown that SDCs do not necessarily act as climate laggards but instead can become climate pioneers or at least followers.

Second, SDCs seem to conceptualise climate change not only as a threat (flooding) but also as an opportunity ('green' jobs). To be able to exploit such opportunities, SDCs have to create cognitive and entrepreneurial pioneership capabilities that require the involvement of local governance actors, businesses, NGOs and citizens. However, to create considerable structural pioneership capacities, SDCs are significantly more reliant on (e.g. financial) support from 'higher' levels of governance such as regional and national governments or the EU (see also Morrison *et al.* 2017).

Third, arguably because of their deep-rooted economic and social problems, SDCs seem relatively willing to endorse transformational climate pioneership styles in the hope of turning their economic fortunes around and improving their poor external image. However, a long term transformational climate pioneership style is contingent on being compatible with core local economic goals, strong support from local governance actors (including local officials, political parties, businesses and societal actors) and at least some support from 'higher' levels of governance.

Fourth, for SDCs there is a tension between learning from each other's best practice in terms of local climate experiments and innovation, and fierce economic competition for inward investment for 'green' jobs (e.g. in the offshore wind energy sector).

Fifth, Liefferink and Wurzel (2017) claim that environmental pioneers and leaders are either 'first in class' or 'best in class' applies to SDCs only to the degree that these cities are more often than not able to show pioneership in their particular place in the urban system, namely structurally disadvantaged maritime port cities. Comparing Bremerhaven and Hull with cities other than SDCs would therefore be akin to comparing apples with oranges. This poses an analytical challenge for the state-focused leader and pioneer typology proposed by Liefferink and Wurzel (2017) because there are arguably more different types of cities than there are different types of states. The applicability of different leadership types – structural, entrepreneurial, cognitive and exemplary – is challenging for assessing urban climate governance although it can add analytical value. Focusing on different types of pioneership (and leadership) creates greater analytical awareness of the fact that SDCs may, for example, try to attract renewable energy industry (structural leadership), build extensive networks and new climate alliances within the city as well as across different governance levels (entrepreneurial leadership) and conceptualise climate change not only as a threat but also as an opportunity when attempting to rebrand the city with the help of innovative local climate innovations (cognitive leadership).

Place clearly makes a difference to how analytical concepts (such as the pioneership/leadership concept) assess local climate governance actions within MLG and polycentric governance structures. The need to make concepts place-specific suggests the importance of contextual factors, which may differ significantly even for the same or similar types of cities such as small-to-medium-sized SDCs. As Le Galés (2002, p.268) has pointed out: 'Each city represents something unique [which is] the result of a unique history'. There are therefore certain path-dependencies, which SDCs with limited resources will find difficult to alter radically. Perhaps the key message is that the analysis of climate change policy in SDCs can be helpful in demonstrating how concepts of environmental leadership originally designed for international and national comparisons can also be applied to the urban scale. Moreover, as much as it sheds light on large-scale trends towards multi-level and polycentric climate governance, urban comparative analysis also helps to expose differences in how climate leadership plays out at the urban scale. Clearly we need additional research on SDC climate governance to build more robust generalisable conclusions.

Notes

1. Hull is formally called Kingston-upon-Hull although it is widely referred to as Hull (see: http://www.hullhistorycentre.org.uk/discover/hull_history_centre/about_us/historyofhull.aspx [Accessed 5.5.2017].
2. It is not yet clear whether the unemployment figures for 2016 constitute an outlier. One reason for the improved employment figure for 2016 is due to the fact that Hull was the UK's City of Culture which resulted in about 800 additional jobs (University of Hull 2018). Other recently created jobs in Hull

are relatively insecure jobs in the gig economy. Table 1 states the unemployment figures for 2015 because they are more typical for the last few decade.

Acknowledgments

Rudi Wurzel and Andrew Jonas are grateful to the British Academy (grant no. SG 131240) and the University of Hull for funding. The authors thank their interviewees. More than 70 interviews with local politicians, officials, businesses and NGOs took place in Bremerhaven and Hull while three were carried out in Berlin and Dessau in 2014–2017. We are grateful to the referees as well as to the journal editors, Anthony Zito and Chris Rootes, for their very helpful comments. We delivered an earlier version of this manuscript at the Innovations in Climate Governance (INOGOV) funded workshop on 'Pioneers and Leaders in Polycentric Climate Governance (PiLePoC)' in Hull on 15–16 September 2016. The usual disclaimer applies.

Disclosure statement

No potential conflict of interest was reported by the authors.

ORCID

Rüdiger K.W. Wurzel ⓘ http://orcid.org/0000-0001-5873-4232

References

AWI, 2009. *Konzeptstudie Klimastadt Bremerhaven*. Bremerhaven: Alfred-Wegener-Institut.
Bertelsmann Stiftung, 2017. *Bremerhaven – pendler – 2013–2015*. http://www.wegweiser-kommune.de/statistik/bremerhaven+pendler+2013-2015+tabelle [Accessed 15 May 2017].
Bulkeley, H. and Betsill, M., 2005. Rethinking sustainable cities: multilevel governance and the 'urban' politics of climate change. *Environmental Politics*, 14 (1), 42–63. doi:10.1080/0964401042000310178
Bulkeley, H., Castán Broto, V., and Edwards, G., 2015. *An urban politics of climate change*. London: Routledge.
Burns, J.M., 1978. *Leadership*. New York: Harper & Row.
Centre for Cities, 2014. Cities Outlook 2014. http://www.centreforcities.org/wp-content/uploads/2014/01/14-01-27-Cities-Outlook-2014.pdf [Accessed 15 May 2017].
Eckersley, P., 2018. Who shapes local climate policy? Unpicking governance arrangements in English and German cities. *Environmental Politics*, 27 (1), 139–160. doi:10.1080/09644016.2017.1380963
Gupta, J. and Grubb, M.J., eds, 2000. *Climate change and European leadership: a sustainable role for Europe?* Dordrecht: Kluwer Publishers.
Homsy, G.C. and Warner, M.E., 2015. Cities and Sustainability: Polycentric Action and Multilevel Governance. *Urban Affairs Review*, 51 (1), 46–73.
Hooghe, L., ed., 1996. *Multi-level governance and European integration*. Oxford: Clarendon Press.

Hull City Council, 2010. *Kingston upon Hull climate change 2010–2020*. Hull: One Hull.
Hull City Council, 2013. Energy City. http://cityplanhull.co.uk/index.php/energy-city-2/[Accessed 06 September.2016].
Hull Data Observatory, 2017. Local economy overview (Hull). http://109.228.11.121/IAS_Live/profiles/profile?profileId=9 [Accessed 10 March 2018].
Hull UK City of Culture, 2017. Somewhere becoming Sea. https://www.hull2017.co.uk/whatson/events/somewhere-becoming-sea/[Accessed 24 May 2017].
Jänicke, M., 2014. Multi-level reinforcement in climate governance. *In*: A. Brunnengräber and M.R. Di Nucci, eds. *Im Hürdenlauf zur Enegiewende*. Berlin: Springer, 35–47.
Jänicke, M. and Wurzel, R., 2019. Leadership and lesson-drawing in the European Union's multilevel climate governance system. *Environmental Politics*, 28, 1.
Jonas, A.E.G., *et al.*, 2017. Climate change, the green economy and re-imagining the city: the case of structurally disadvantaged European maritime port cities. *Die Erde*, 148 (4), 197–211.
Jonas, A.E.G., Gibbs, D., and While, A., 2011. The new urban politics as a politics of carbon control. *Urban Studies*, 48, 2537–2544.
Jordan, A., *et al.*, eds, 2018. *Governing climate change: polycentricity in action*. Cambridge: Cambridge University Press.
Kaelble, H., 1988. *Vers un societé européene, 1880–1990*. Paris: Belin.
Kane, R.J., 2005. Compromised police legitimacy as a predictor of crime in structurally disadvantaged communities. *Criminology*, 43 (2), 462–498. doi:10.1111/j.0011-1348.2005.00014.x
Kemmerzell, J., 2017. Überlokales Handeln in der lokalen Klimapolitik. *In*: M. Barbehön and S. Münch, eds. *Variationen des Städtischen - variationen lokaler Politik*. Wiesbaden: Springer, 245–271.
Kern, K., 2019. Cities as leaders in EU multi-level climate governance? Embedded upscaling of local experiments in Europe. *Environmental Politics*, 28, 1.
Kern, K. and Bulkeley, H., 2009. Cities, Europeanization and multi-level governance. *Journal of Common Markets Studies*, 47 (1), 309–332. doi:10.1111/j.1468-5965.2009.00806.x
Kythreotis, A.P. and Bristow, G.I., 2017. The 'resilience trap': exploring the practical utility of resilience for climate change adaptation in UK city-regions. *Regional Studies*, 51 (10), 1530–1541. doi:10.1080/00343404.2016.1200719
Le Galès, P., 2002. European cities. In: *Social conflicts and governance*. Oxford: Oxford University Press.
Liefferink, D. and Wurzel, R.K.W., 2017. Environmental leaders and pioneers: agents of change. *Journal of European Public Policy*, 24 (7), 651–668. doi:10.1080/13501763.2016.1161657
Mederake, L., 2015. *Opportunities for the local government of Bremerhaven provided by the project "Klimastadt Bremerhaven" in times of limited municipal room for maneuver*. Bremen: Hochschule Bremen.
Migration Observatory, 2014. Changes to the migrant population of Yorkshire and the humber 2001–2011. http://www.migrationobservatory.ox.ac.uk/press/changes-to-the-migrant-population-of-yorkshire-and-the-humber-2001-2011/ [Accessed 8 March 2018].
Migration Yorkshire, 2018. Hull's newcomers in 2017 http://www.migrationyorkshire.org.uk/?page=hullnewcomers [Accessed 17 August 2018]
Mol, A., 2010. Environmental authorities and biofuel controversies. *Environmental Politics*, 19 (1), 61–79. doi:10.1080/09644010903396085

Morrison, T.H., et al. 2017. Mitigation and adaptation in polycentric systems: sources of power in the pursuit of collective goals. *WIREs Climate Change*, 7, 1–16.

Ostrom, E., 2009. *A polycentric approach for coping with climate change*, Policy Research Working Paper 5095. Washington: World Bank.

Ostrom, E., 2014. A polycentric approach for coping with climate change. *Annals of Economics and Finance*, 15 (1), 97–134.

Porter, M., 1990. The competitive advantage of nations. *Harvard Business Review*, March/ April,73–93.

Röhl, K.-H. and Schröder, C., 2017. Regionale Armut in Deutschland. Risikogruppen erkennen, Politik neu ausrichten. IW-Analysen Nr. 113. Cologne: Institut der deutschen Wirtschaft.

Schreurs, M. and Tiberghien, Y., 2007. Multi-level reinforcement: explaining European Union leadership in climate change mitigation. *Global Environmental Politics*, 7 (4), 19–46. doi:10.1162/glep.2007.7.4.19

Singleton, B.E., 2018. What is missing from Ostrom? Combining design principles with the theory of sociocultural viability. *Environmental Politics*, 26 (6), 994–1014. doi:10.1080/09644016.2017.1364150

Statistisches Landesamt Bremen, 2017. *Zahlenspiegel*. November 2017. https://www.statistik.bremen.de/detail.php?gsid=bremen65.c.2076.de

Strange, S., 1998. *States and markets*. London: Pinter Publishers.

SUBV, 2010. *Klimaschutz- und Energieprogramm 2020*. Bremen: Der Senator für Umwelt, Bau und Verkehr (SUBV).

University of Hull, 2018. University of Hull reveals UK city of culture evaluation. https://www.hull.ac.uk/work-with-us/more/media-centre/news/2018/city-of-culture-evaluation.aspx [Accessed 21 August 2018].

Wegweiser Kommune, 2016. Typ 9: Stark schrumpfende Kommunen mit besonderem Anpassungsdruck. http://www.wegweiser-kommune.de/documents/10184/33037/Demographietyp+9.pdf/6f204283-d2da-4ab4-8065-672bd211aa20 [Accessed 3.4.2018]

While, A., Jonas, A.E.G., and Gibbs, D., 2010. From sustainable development to carbon control: the eco-restructuring of the state and the politics of urban and regional development. *Transactions Institute of British Geographers*, 35 (1), 76–93. doi:10.1111/j.1475-5661.2009.00362.x

Wilson, W.J., 1987. *The truly disadvantaged*. Chicago: University of Chicago Press.

Wurzel, R., Connelly, J., and Liefferink, D., eds, 2017. *The European union in international climate change politics*. London: Routledge.

Follow the leader? Conceptualising the relationship between leaders and followers in polycentric climate governance

Diarmuid Torney

ABSTRACT
Existing scholarship on climate governance has not sufficiently considered the relationship between climate leaders/pioneers and followers. Because of the global commons nature of climate change, unilateral leadership or pioneership by one or a small number of actors will be insufficient to combat climate change effectively. The need to take seriously the relationship between leaders and followers is all the greater in the wake of the 2015 Paris Agreement, which emphasises diffuse, bottom–up action. The relationship between leaders and followers in polycentric climate governance is unpacked in this contribution. What types of actors can be climate followers? Through what pathways can followership emerge and how can we capture the essential characteristics of leader–follower relationships? What conditions facilitate (or hinder) followership? The utility of the approach is illustrated using cases of EU climate leadership and (non-) followership of other actors.

Introduction

Scholars have expended significant effort to identify and conceptualise patterns of leadership in global climate change politics. However, most literature on leadership and pioneership in climate governance has focused relatively little on the role of climate followers and the leader–follower relationship. A recent overview of political leadership studies (Blondel 2014) highlighted the relationship between leaders and followers as one of three priorities for future research. There are exceptions, such as the studies by Parker, Karlsson and colleagues (see e.g. Karlsson *et al.* 2012, Parker *et al.* 2012, Parker and Karlsson 2018) that provide rich data on patterns of followership in global climate negotiations. I have previously sought to conceptualise the leader–follower relationship as the result of three factors:

the drivers of engagement of the leader with would-be followers, the form of that engagement, and the response of followers (Torney 2015a, 2015b). Here, I build on that conceptualisation but deepen it significantly.

To varying degrees, research on policy diffusion, convergence, transfer, and learning has addressed questions about how and whether policies, institutions and norms spread from their original source. Policy diffusion research has typically examined patterns of policy development among large populations of cases (e.g. Tews *et al.* 2003). Studies of policy transfer and learning have focused more on the detailed and specific processes through which policies spread from one jurisdiction to another (e.g. Rose 1993, Dolowitz and Marsh 1996). While that literature provides useful insights, I instead use the concepts of leadership and followership for two reasons. First, research on climate governance has, for some time, used the term leadership. Second, the remit of my analysis is broader than internal policies/arrangements and includes dynamics of international negotiations.

The neglect of followership in the leadership literature is to some extent understandable due to methodological and evidential challenges associated with convincingly identifying followers and followership. For example, how do we know that the actions of a claimed leader or pioneer in fact triggered a change of approach on the part of a would-be follower? The policy transfer and learning literature has also faced this evidential challenge (Rose 1993, Dolowitz and Marsh 1996). As Bennett (1991, p. 231) succinctly puts it, 'the analyst must avoid the pitfall of inferring from a transnational similarity of public policy that a transnational explanation must be at work'. This holds equally for students of followership.

Here I develop a more nuanced understanding of the relationship between leaders and followers in polycentric climate governance. My starting point is Liefferink and Wurzel's (2017) framework. They make a number of important analytical distinctions. First, they distinguish between *leaders* and *pioneers*; leaders usually actively seek to attract followers; pioneers normally do not. Second, they distinguish between four leadership *types*: structural, entrepreneurial, cognitive and exemplary. Third, they distinguish between two leadership *styles*: transactional/humdrum and transformational/heroic. Notably, however, their article 'neither focuses on the interrelations between leaders and followers nor does it assess how leaders and pioneers are perceived by third states' (Liefferink and Wurzel 2017, p. 3). I build on their work here by focusing on this understudied aspect.

Polycentric governance settings arguably present a hard case for the emergence of leader–follower relationships, because they involve many centres of decision-making that are formally independent of each other (Ostrom 2010, 2012). Pushed to its logical extreme, polycentricity entails non-hierarchical governance arrangements in which it may be difficult to distinguish leaders from followers (Liefferink and Wurzel 2018).

Nonetheless, experimentation and learning are important features of polycentric governance systems. Indeed, for Ostrom (2012, p. 365), this is the great promise of a polycentric approach, which 'encourage[s] experimentation and learning from diverse policies adopted by multiple scales'. This focus on learning is echoed in other work on polycentric climate governance (e.g. Rayner and Jordan 2013, Dorsch and Flachsland 2017), though Jordan et al. (2015) question the optimism in some of the literature.

We can distinguish between actors' formal autonomy and their interconnectedness. Although actors in polycentric governance systems may indeed enjoy high levels of formal autonomy, they may nonetheless be interconnected in significant ways with other actors within the system, providing conditions for leader–follower relations to emerge. Moreover, the absence of strong hierarchical mechanisms to secure adequate responses to climate change heighten the importance of pioneers and leaders on the one hand, but also followers – since the global nature of climate change means that one or a small number of actors are unlikely to solve the problem alone.

This is particularly the case in the context of the Paris Agreement, which marked a shift from the top–down global climate governance architecture that characterised the Kyoto Protocol towards a bottom–up approach that grants greater autonomy to states to determine their level of contributions. Paris also sought to capture the diffuse set of climate actions by a variety of non-state actors including cities, municipalities and businesses through the so-called Lima-Paris Action Agenda. Sabel and Victor (2016) argue that the Paris architecture should be taken a step further to allow for greater experimentation and learning among this diverse universe of public and private governance actors.

I define climate followership as the adoption of a policy, idea, institution, approach, or technique for responding to climate change by one actor by subsequent reference to its previous adoption by another actor. Note that there must be intentionality on the part of the follower but not the leader/pioneer. The follower must intentionally follow what a leader pioneer has done. By contrast and as noted above, while a leader will proactively seek to attract followers, a pioneer may not. The range of items that can be the focus of followership is deliberately broad and encompassing because, in keeping with the focus of this volume, a variety of types of actors can be both leaders and followers.

Building on and extending the literature on climate leadership, I develop a framework that answers three questions in respect of the leader–follower relationship. First, who follows? As the other contributions to this volume illustrate well, a diverse range of actors can be climate leaders and pioneers, including not just states and supranational entities such as the EU, but also businesses, NGOs, unions, epistemic communities and individuals. I argue that, in principle, actors in each of these categories can also be followers. However, it is most likely that leader–follower relationships will emerge between actors of the same type, though such dynamics may also emerge between actors across different categories, for example, between the EU (a supranational entity) and a state.

Second, through which pathways can followership emerge, and how can we capture the key characteristics of leader–follower relationships? Following March and Olsen (1998), I distinguish between pathways stemming from a logic of consequences and a logic of appropriateness. On the one hand, leaders may attract followers through coercion or the provision of incentives (logic of consequences). Here we can, in turn, distinguish between involuntary and voluntary followership. On the other hand, leaders may attract followers through providing model or exemplary performance (i.e. exemplary leadership/pioneership), or through provision of knowledge that is considered particularly appropriate or relevant (logic of appropriateness).

Third, what conditions facilitate or hinder followership? Here, we can distinguish again between a logic of consequences and a logic of appropriateness. In the case of a logic of consequences, the degree of followership is likely to depend on the type and extent of power resources at the disposal of the leader or pioneer, as well as the degree of power asymmetry between the leader/pioneer and the would-be follower. In the case of a logic of appropriateness, the perceived legitimacy of the example or knowledge being promulgated by the leaders will affect the degree of followership, as will the ability of the purported leader to frame that knowledge in ways that resonate with the norms and interests of the would-be follower. The characteristics of the would-be follower are also likely to be important, including for example whether interest groups are mobilising for stronger climate action (in the case of state actors).

I structure the analysis around these three questions. Subsequently, I use the case of EU climate leadership and the followership of other actors to illustrate the utility of this framework. Considering the novelty of the followership research, the case study is explorative, and, since the case is limited to public governance institutions, it cannot assess the full diversity of the framework developed below. Future research may take up that task.

Who can be a climate follower?

Followers can emerge in response to leadership on the part of actors in a diverse set of categories. In the category of public governance actors, states as well as supranational (e.g. the EU) and subnational (cities, municipalities) entities can be followers. For example, while observers commonly portray the EU as a climate and environmental leader, it was not always thus. In the early days of environmental awareness, the EU followed the leadership of other international actors, including the United States (Meyer 2011). Similarly, cities and municipalities can be followers as well as leaders, though as Kern (2019 – this volume) notes, many smaller European cities and towns have not in fact followed the leadership of larger cities.

Equally, private actors can also be followers as well as leaders. Networks such as the Carbon Disclosure Project (2016) provide mechanisms whereby follower businesses can adopt public disclosure practices of early adopters. Businesses may also follow the lead of other businesses or social enterprises (Biedenkopf *et al.* 2019 – this volume). Similarly, non-governmental organisations (NGOs) and other civil society groups including trade unions can be followers as well as leaders. For example, different national branches of Friends of the Earth Europe advocated for national climate laws, following the lead of Friends of the Earth England, Wales and Northern Ireland that had earlier and successfully lobbied for the introduction of the UK Climate Change Act (Torney 2017, Carter and Childs 2018).

The individual level is more ambiguous with respect to followership. Individuals can, of course, be followers, particularly as concerned citizens or responsible consumers. However, conceptualising followership in this way risks casting the net too wide, since we would end up potentially dealing with entire populations. If we narrow the focus to 'notable' individuals considered by Liefferink and Wurzel's (2017) – such as scientists and public intellectuals – it seems less likely that such individuals could be followers as well as leaders.

Leader–follower dynamics are most likely to materialise among actors within one of the above categories, both because actors are more likely to search within their own peer group for examples of leadership, and because like actors are likely to face similar challenges. In principle, it may be possible for a leader in one actor category to attract followers from another category, but different theoretical perspectives provide different accounts of its likelihood. For state-centric perspectives, only states can form analytically meaningful leader–follower relationships, not least because, according to this view, states are the only meaningful governance actors. Multilevel governance perspectives allow for followers to emerge at levels of governance different from leaders within the domain of public actors. For example, state actors may follow the lead of the supranational EU, which has adopted a wide range of EU-wide climate policy measures. Similarly, subnational entities may follow the lead of states in other parts of the world, or vice versa. Only polycentric governance perspectives allow for leader–follower relationships among all actor categories identified above, including those spanning public and private actor domains.

As well as distinguishing between followers according to their actor type, we can also distinguish between followers according to *what* they follow. Wurzel *et al.* (2019 – this volume) distinguish between the internal and external ambitions of leaders, arguing that leaders with low internal but high external ambition can be classified as *symbolic* leaders whereas leaders with both high internal and external ambition can be classified as *substantive* leaders. We can make an analogous distinction with respect to followers. Followers can follow leaders' high external ambition – their declared commitment – for example by committing to strong decarbonisation targets. Followers can also follow leaders' high internal ambition by implementing concrete measures to achieve targets.

Through what pathways can climate followership materialise?

How can we conceptualise the pathways through which followership may materialise, and the connections between followership and leadership/pioneership? It is helpful to distinguish between followership based on a logic of consequences and followership based on a logic of appropriateness (March and Olsen 1998). A variety of different possible followership pathways can be identified, which can be mapped onto the different leadership *types* set out by Liefferink and Wurzel (2017).

Drawing on a logic of consequences, followers may be induced to follow through the use of structural power by a leader to induce followership. Followership of this kind will emerge as a response to structural leadership. Not all forms of power are likely to be useful for inducing followership. For example, military power (in the case of state actors) is unlikely to be helpful for a climate leader seeking to attract followers. By contrast, economic power, such as granting or denying market access, is more likely to attract climate followers. This followership pathway (and leadership) is more likely to arise in respect of public governance actors such as states as well as supranational and subnational entities, but is also possible in the case of business actors. It is less likely, however, in the case of civil society groups and individuals.

Echoing the policy convergence and transfer literature (e.g. Bennett 1991, Dolowitz and Marsh 1996), we can distinguish between involuntary followership, involving coercion on the part of the leader/pioneer, and voluntary followership, involving provision of incentives by the leader/pioneer. Followers may effectively have no other choice if they are coerced into following by a structural leader exerting material power. For example, the accession process for central and eastern European states provided the EU with strong leverage to bring those countries' environmental policies into line with existing EU laws (Carmin and VanDeveer 2004; see also Schimmelfennig and Sedelmeier 2004). Followers may have a choice but may opt to follow primarily as a result of being provided material incentives that alter their cost–benefit calculations (Damro 2012, Postnikov 2018). This distinction between involuntary and voluntary followership is not binary, but rather we should see it as a spectrum of possible followership pathways.

It is also possible for followership to emerge in response to structural leadership exercised by a pioneer rather than a leader, that is, by an actor that does not seek to attract followers. Through the external effects of, for example, market power, a pioneer may induce followership without explicitly intending to do so. This could occur if a large market raises climate or environmental standards for goods and services, and applies these to both domestic and foreign producers in order to create a level playing field. Countries wishing to export to the large market may raise their own domestic standards in order to comply with regulatory frameworks in the larger market (Vogel 1995).

Turning to a logic of appropriateness, followership can materialise through learning and persuasion, including from new ideas promulgated by exemplary or cognitive leaders. In such cases, followers follow leaders not because they are coerced or incentivised to do so, but because they believe the models provided by a leader are superior in some way and worthy of followership. In contrast to a logic of consequences, followership of this kind is not restricted to particular kinds of actors.

Followership can materialise as a result of learning from exemplary leadership. In this case, followership is the result of actors viewing a particular approach to a policy problem as particularly effective or innovative (see Jänicke and Wurzel 2019 – this volume). According to Liefferink and Wurzel (2017), cognitive leadership involves efforts through persuasion to redefine ideas that constitute actors' interests. Followership in this context occurs where other actors accept the leader's (re)framing of an issue. Importantly, different types of leadership – and by extension followership – play out over different timescales. In particular, cognitive leadership usually takes longer, since it may take time to transform other actors' existing conceptions of their interests and causal beliefs. Turning to the distinction between leaders and pioneers, it is equally possible for exemplary pioneers to attract followers. It is less likely for pioneers to attract followers through cognitive pioneership, since such pioneership involves a degree of intentionality on the part of the pioneer.

Entrepreneurial leadership cuts across the distinction between logics of social interaction. As Liefferink and Wurzel (2017) note, entrepreneurial leadership is, in practice, rarely exercised in isolation. Rather, it is more often used in support of other kinds of leadership. Therefore, it can be difficult to conceptualise followership as a response to entrepreneurial leadership independent from other types of leadership. For example, we can see voluntary followership resulting from a combination of entrepreneurial and exemplary leadership in the remarkable similarities seen across parties' Intended Nationally Determined Contributions (INDCs) submitted prior to the 2015 Paris climate conference. Much of this was the result of both entrepreneurial and exemplary leadership on the part of the EU and its member states. By publishing its own INDCs at an early stage in the process, the EU provided exemplary leadership. In parallel, the EU and its member states, along with other external actors including the US, provided extensive support to over 100 countries to enable them to publish similar INDCs, thereby employing entrepreneurial leadership (Dupont et al. 2018). It was this combination of exemplary and entrepreneurial leadership, rather than either on its own, that resulted in followership on the part of many other states.

What conditions facilitate followership?

When considering the conditions under which followership is likely to emerge, we can distinguish between explanatory factors specific to the (would-be) leader and those specific to the (would-be) follower. What characteristics of a leader are likely to attract followers? It is helpful again to distinguish between leadership based on a logic of consequences versus leadership based on a logic of appropriateness, since relevant factors will differ across these two categories.

Turning first to a logic of consequences, the nature of a leader's power resources as well as their ability to mobilise those resources is important, as is the relevance of the source of power to the issue at stake. For example, military power may not be conducive to securing climate followership. Recall, also, that this category of leadership is most applicable to public governance actors, though it may also apply to businesses. Success in attracting followers is likely to be determined significantly by the degree to which an aspiring leader mobilises the requisite material resources to exercise structural leadership over other actors. The greater the issue-specific power resources of the leader, e.g. the greater their contribution to the environmental problem in question, the more likely they are to be able to exercise structural power. This is dependent on both the presence of such resources, but also on decision-making procedures and the priority attached to climate and environmental affairs. If other aspects of the leader's relationship are prioritised, for example, it may be that they will be unable or unwilling to mobilise resources at the service of environmental policy goals.

In the case of a logic of appropriateness, what characteristics of a leader are likely to result in followership? In the case of an exemplary leader seeking to provide lessons to others, the perceived legitimacy of the leader is likely to be important. How genuinely innovative or exemplary do the followers consider the model being provided by the leader to be? Does the leader 'walk the walk' or is its leadership perceived to be merely symbolic, and are the models and knowledge provided by the purported leader viewed as authoritative and legitimate (Parker *et al.* 2012)? In the language of Wurzel *et al.* (2019 – this volume), is the leader a *symbolic* or a *substantive* leader? Other things being equal, a substantive leader is more likely to attract followers. This is less dependent on purposive actions of the leader to attract followers, and can apply to pioneers who have little or no desire explicitly to attract followers as well as to leaders. Furthermore, followership is likely to emerge the more the model the leader adopts is realistic for, and fits, the characteristics and circumstances of the follower. There needs to be a fit between what the leader offers and what the follower wants to and can accept.

The greater the 'normative gap' between the leader and follower, the less likely it is for followership to emerge (Torney 2015b). However, a cognitive leader is more likely to attract followers through persuasion if they can successfully build congruence between a new approach or reframing of a policy problem and pre-existing norms and beliefs of other actors are important (Cortell and Davis 2000, Checkel 2005). It is also dependent on the capacity to disseminate those concepts, framings, and new knowledge to those other actors. It depends partly on the ability to produce new knowledge and understanding, but this is not enough if the leader cannot disseminate it to prospective followers. An entrepreneurial leader, finally, is more likely to attract followers the more it builds up knowledge of the interests of other relevant actors. Being able to design complex package deals that bring all actors on board is crucially dependent on understanding what the interests and preferences of other actors are, as well as possessing the creativity to design those package deals.

Finally, the level of ambition of the leader is likely to matter, but this leadership characteristic generates conflicting expectations. Transformational leadership, because it is likely to be seen as more ambitious and potentially more legitimate, may be more likely to attract followers. However, a leader who aims at transformational change may seek to push other actors further beyond their existing interests and preferences, thereby making followership less likely. In these circumstances, transactional leadership may be more likely to resonate with the interests of other actors.

What characteristics of a potential follower are likely to facilitate followership? In respect of states in particular, domestic characteristics and circumstances of the would-be follower will also shape the degree of followership, including (in the case of states) the openness of the political system of the follower to engagement by external actors (Risse-Kappen 1994, Checkel 2005). Again in the case of states, the balance of domestic interests between supportive and opposing interests will have a bearing on the likelihood of followership. In the case of followership of a structural leader, the nature of the power relationship is important. Highly asymmetrical power relations, in which the would-be follower effectively has no other choice, are likely to enable the leader to coerce the follower. If the leader tries to induce followership through provision of incentives rather than coercion, the dependence of the follower on the leader, and their alternative options, will matter also.

Table 1 summarises the framework developed so far. It identifies who can be a climate follower, pathways to followership, and the factors that enable followership. In the next section I consider the case of the EU in global climate governance, using examples from the EU's efforts to secure followership to the extent to which this illustrates the utility of this framework.

Followership in practice: EU leadership and (non-)followership

Scholarly research on EU climate leadership (e.g. Vogler 1999, Gupta and Grubb 2000, Schreurs and Tiberghien 2007, Oberthür and Roche Kelly 2008) has had less to say on whether the EU has successfully attracted followers, though there are exceptions (Kilian and Elgström 2010, Lindenthal 2014, Torney 2015a, 2015b, Parker and Karlsson 2017, Parker et al. 2017). I use the case of EU leadership and (non-)followership by other actors to illustrate the utility of the analytical framework developed above, though due to space limitations I do not seek to test the framework systematically. This case study is restrictive in two respects. First, the EU has been a *leader* rather than a *pioneer*. Second, the case of the EU is 'state-centric', even though the EU is not a state in the traditional sense. As such, it cannot fully exploit the first dimension of the framework, namely the question of who follows. It is more instructive with respect to the second and third

Table 1. Dimensions of followership.

Who can follow?	Pathways of followership	Facilitating conditions
Supranational actors	*Logic of consequences:*	*Leader characteristics:*
States		Extent and type of material power
Subnational actors	Coercion	Ability and willingness to mobilise resources
Businesses	(involuntary)	
Trade unions	Incentives	Legitimacy and credibility
NGOs	(voluntary)	Ability to (re)frame issues
Scientists	*Logic of appropriateness:*	*Follower characteristics:*
Within category dyads (e.g. state-state, business-business, etc.) more likely		Domestic political structure
	Learning	Pre-existing norms and ideas
	Persuasion	*Relationship characteristics:*
		Symmetry/asymmetry of power relationship
		Size of normative gap between leader and potential follower

Source: Author's compilation

dimensions: pathways of followership and conditions that facilitate (or constrain) followership. My following assessment is structured according to the four followership pathways: *coercion; incentives; learning;* and *persuasion*. For each pathway, in the case discussion I highlight factors that either facilitate or constrain followership.

Coercion

Coercion can bring about involuntary followership. We can disaggregate it into physical and legal coercion. The former is not relevant in the current context. In an EU context, coercion resonates with Damro's (2012) conceptualisation of 'Market Power Europe'. To what extent and under what conditions has the EU succeeded in attracting climate followership through coercion?

The case of international aviation and the EU Emissions Trading Scheme (ETS) illustrates the limited ability of the EU to use coercion in the face of strong and united opposition. The EU attempted to exercise structural leadership by seeking to include international aviation in the EU ETS. However, a range of countries including the United States, Russia, China, and India responded angrily. In particular, they claimed the EU move was the thin end of the wedge – the first of potentially many such unilateral border measures imposed by the EU. In the face of this strong opposition, the EU backed down on its proposal in favour of negotiations within the framework of ICAO. In this case, the power distribution (lack of asymmetry in favour of the EU) combined with a lack of willingness on the part of the EU and its member states to mobilise resources in favour of its climate goals helps to explain the failure to attract followers.

To some extent a process of Europeanisation of national climate and environmental policies across EU member states speaks to the EU's ability to secure climate followership through coercion. However, the 2014 European Council decision on a post-2020 climate and energy framework and ongoing negotiations over the Energy Union point to a partial roll-back of climate and energy policy (European Council 2014). The move from nationally binding renewable energy targets under the 2020 package to targets that are binding at EU level only in the 2030 framework illustrates this trend. Moreover, some newer member states increasingly sought to undermine existing EU climate and energy policies as well as seeking to lower ambition of future EU climate targets (Skovgaard 2013). Poland in particular has been central to opposition to increasing the level of ambition of EU climate policies.

Incentives

The provision of material incentives can attract voluntary followers though, as discussed above, the extent to which such followership is truly voluntary is a question of degree, depending in part on the nature of the power relationship and whether the prospective follower has realistic alternatives. To what extent has the EU succeeded in attracting followers through provision of incentives? In one of the more successful instances of EU climate leadership, the EU secured entry into force of the Kyoto Protocol through a deal under which Russia ratified the Protocol in return for EU support for Russian accession to the World Trade Organization (WTO) (Damro 2006). The EU did not have a significant material power advantage over Russia, another major power, though arguably the EU did possess considerable leverage due to Russia's desire to join the WTO. A significant factor here was the EU's willingness to mobilise resources in support of securing climate followership, and in particular to use another policy field (trade) to secure climate followership. Indeed, it was only Russian and Japanese followership that made the EU's leadership successful.

More broadly, the EU's inclusion of climate and environment provisions in bilateral and regional trade agreements shows both the potential and the limitations of the EU's attempts to attract followers through provision of incentives. Postnikov (2018) finds that the EU has increasingly included climate and environment provisions in trade agreements with third countries and regions. Moreover, not only has the scope of such provisions expanded over time, but environmental standards in trade agreements have also been made more legally binding, putting them on the same footing as trade issues. However, he also notes that the EU's approach to enforcement of such climate and environmental standards remains 'soft', particularly by comparison with the approach adopted by the US, thereby limiting its effectiveness. These findings further reinforce the importance of the willingness to mobilise resources in support of climate and environmental goals in determining ability to attract followers through provision of incentives.

The EU has also used provision of climate finance and capacity-building support to developing countries as a means of attracting followers. The EU and its member states contributed €20.2 billion in climate finance to developing countries in 2016, and have contributed approximately half of all funding to the Global Environment Facility and the Green Climate Fund (Council of the European Union 2017, Dupont *et al.* 2018). The EU and several member states also engaged in capacity-building in about 100 countries to support elaboration of INDCs in advance of COP21. Most obviously, the provision of these incentives resulted in many third countries around the world following the EU in framing meaningful INDCs, though of course the EU was not acting alone (Dupont *et al.* 2018).

Learning

How successful has the EU been in attracting followership through a learning pathway? It can be hard to conclusively prove causality in cases of learning, not least because there may be other governance actors promoting the same policy approach. One case that illustrates these complexities well is emissions trading. While the EU has made its EU ETS a centrepiece of its response to climate change and has sought to promote its use in other jurisdictions, the idea of emissions trading was not European in origin. Rather, it was developed in an American context as a response to the problem of sulphur dioxide emissions (Biedenkopf 2012). Moreover, in specific instances where the EU has sought to promote emissions trading beyond its borders – most notably in China – the EU is one among several actors seeking to do so. In the Chinese case, a range of other actors including the US, Australia and the World Bank have also actively promoted carbon emissions trading (Torney and Gippner 2018).

Nonetheless, we can reasonably argue that the EU ETS served as a reference point for Chinese policymakers seeking to develop emissions trading in China. The European Commission along with the UK and Germany have been actively involved in supporting the Chinese government's attempts to build emissions trading pilots (Biedenkopf and Torney 2014). However, interestingly, evidence suggests that the Chinese government has learned as much what not to do as what to do from the EU ETS (Torney and Gippner 2018). The fact that the EU has the deepest and most extensive experience in developing carbon emissions trading made it a natural partner for the Chinese government on emissions trading, but perceived weaknesses in the EU ETS design limited the credibility of the EU model. While the literature on policy transfer and learning allows for both negative as well as positive lesson-drawing (Rose 1993, Dolowitz and Marsh 1996), it is less clear that negative lesson-drawing such as this constitutes followership.

We can observe an example of the EU's failure to attract followers through exemplary leadership in its approach to setting conditional emissions reduction targets in advance of the 2009 Copenhagen climate conference (COP15). Well in advance of COP15, the European Council in March 2007 agreed to a headline goal of a 20% reduction in greenhouse gas emissions by 2020 (relative to 1990 levels) as part of the EU's contribution to a post-2012 global climate agreement. This target was to be raised to 30% in the context of a global agreement in which 'other developed countries commit themselves to comparable emission reductions and economically more advanced developing countries to contributing adequately according to their responsibilities and respective capabilities' (European Council 2007, p. 12). While other major emitters did indeed announce pledges of 2020 climate action in the context of the Copenhagen negotiations, their pledges were not in line with the European Council stipulations of 2007. In this case, a large part of the explanation for the EU's failure to attract followers rests in the disconnect between what the EU was proposing and pre-existing normative frames and conceptions of material interest in other key states – what I have previously termed the 'normative gap' in the EU's external climate relations (Torney 2015b, chapter 4; see also van Schaik and Schunz 2012).

Persuasion

Cognitive leaders can seek to attract followers through persuasion. However, again, conclusively proving causality is difficult in the case of a persuasion pathway, not only because of the possibility of multiple nodes of leadership beyond the actor under investigation but also because of the long time horizon over which cognitive leadership and persuasion mechanisms often play out.

The EU (and its member states) arguably both successfully and unsuccessfully attracted followers in reframing climate change as an opportunity and not just a threat. Successive German governments from the late 1990s onwards framed environmental policy in terms of 'ecological modernization' and 'ecological industrial policy' to emphasise opportunities for innovation (Jänicke 2010, p. 134). Similarly in the UK, Prime Minister Tony Blair highlighted business opportunities associated with combating climate change (Rayner and Jordan 2010, p. 102). This reconceptualisation also featured in EU Council conclusions, which, from 2000 onwards, began to highlight economic benefits associated with climate action (Oberthür and Dupont 2010, pp. 86–87). Of course, actors outside the EU, including IGOs such as the World Bank and IMF, pushed such reframing as well. Nonetheless, it is reasonable to assume that the global diffusion of such reframing would have been slower or less pronounced in the absence of the EU. A long-term commitment to this persuasion as well as its internal credibility within the EU underpinned the ability of the EU – albeit partial and over a relatively long time horizon – to attract followers through persuasion. Progressive efforts to decouple emissions from growth within the EU underpinned the persuasiveness of the EU's arguments, as did high profile examples such as Germany's *Energiewende*.

In a different context, we can also see followership through the persuasion pathway in the EU's successful mobilisation of coalitions of support within the UNFCCC process at the Durban (2011) and Paris (2015) climate conferences. The EU's leadership in these contexts is best characterised as entrepreneurial which, as I note above, cuts across the logics of social interaction. The EU's mobilisation of the Durban coalition as well as the High Ambition Coalition during COP21 was underpinned by provision of incentives, including climate finance and capacity building support. However, the EU also engaged in persuasion and successfully attracted followers, particularly among developing countries (Torney and Cross 2018). Both instances are particularly interesting from a followership perspective because developing countries had the choice of leaders to follow – an EU-led coalition or an alternative coalition led by China and India.

Conclusions

I have here sought to fill a gap in the existing literature on climate and environmental leadership and pioneership. To date, this scholarship has not sufficiently considered the relationship between leaders/pioneers and followers. Taking the climate and environmental leadership literature as a point of departure, I posed three sets of questions: Who follows? Through what pathways can followership materialise? What conditions facilitate followership?

In response to the first question, in principle a wide range of actors can be followers, including states, supranational and subnational entities, businesses, NGOs, unions, epistemic communities of scientists and experts, and individuals. Within-category dyadic leader–follower relationships (e.g. state–state, business–business) are more likely, though some across-category dyadic relationships are possible. In relation to the second question, I have posited that followership can materialise through four principal pathways, which in turn can be characterised by either a logic of consequences or a logic of appropriateness. In the former case, followership can emerge as a result of either coercion, which is involuntary, or provision of incentives, which is voluntary to a greater or lesser extent. In the latter case, followership can result from learning or persuasion.

Third, the conditions that facilitate (or constrain) followership vary according to which logic of social interaction is in play, and concern characteristics of the leader and potential follower as well as their interrelationship. In the case of a logic of consequences, the extent and type of material power will matter significantly, as well the leader's ability and willingness to mobilise resources in support of its climate leadership. In the case of a logic of appropriateness, the perceived legitimacy and credibility of the leader as well as their ability to (re)frame issues in targeted ways is likely to matter significantly. The follower's domestic political structure (in the case of state actors) as well as their pre-existing norms and interests will also have a significant bearing on the propensity for followership to emerge.

I assessed the analytical framework using examples from the EU's climate leadership. These cases do not test the framework in a rigorous way, but rather were intended to demonstrate its utility. Because of the nature of the EU, the case study was limited in respect of the first question to considering public actors. In respect of the second question, leader–follower relations have tended to emerge more often through incentives, learning and persuasion pathways than through coercion, echoing previous work on EU external relations (Youngs 2001). The third question concerns facilitating conditions. With respect to a logic of consequences, the EU's attempts to include international aviation in the EU ETS show that the willingness of the leader to mobilise resources in support of attempted coercion, particularly in the face of strong opposition, matters, as does the pre-existing preferences of the would-be followers. The EU has had more success in attracting followers through provision of incentives, as illustrated by the EU's deal with Russia regarding Kyoto Protocol ratification. In contrast to the coercion pathway, in the case of incentives the EU's requests of would-be followers have been more in line with preferences of would-be followers. With respect to a logic of appropriateness and the EU's ability to attract followers through learning, the EU ETS provides mixed evidence. China in particular has looked to the EU's example in constructing its own ETS due to the EU's exemplary role in this area. However, the difficulties faced by the EU ETS have raised questions over the credibility of the EU's model. Finally, both the EU's ability to frame its own experiences for international audiences as well as the extent of the normative gap between the EU and third countries have shaped the EU's ability to attract followers through persuasion.

The illustrative case of EU leadership and followership has only scratched the surface empirically, and future research is necessary to further test the propositions I have developed here. Three avenues for future research deserve particular attention. First, as noted above the EU examples concerned a narrow range of types of actors and thus could not really address the first element of the framework: who can be a follower. Future research should test the framework against a wider variety of types of actors. Second and related, future studies could flesh out the third element of the framework particularly in the case of non-state actors. Third, in an increasingly polycentric post-Paris climate governance context, future state-based research on followership ought to broaden the empirical focus beyond the consideration here of the EU as a climate leader. In particular, as China assumes an increasingly prominent role in global climate governance, it will be important to pose similar questions of China.

Acknowledgements

I thank Duncan Liefferink, Rüdiger Wurzel, three anonymous referees for helpful comments on earlier drafts, and Louise FitzGerald for editorial assistance. I am grateful for helpful comments and suggestions to participants in a workshop at the University of Hull in September 2016 on 'Pioneers and Leaders in Polycentric Climate Governance', organised in the framework of the INOGOV Cost Action Network. The usual disclaimer applies.

Disclosure statement

No potential conflict of interest was reported by the author.

ORCID

Diarmuid Torney http://orcid.org/0000-0003-4156-9044

References

Bennett, C.J., 1991. Review article: what is policy convergence and what causes it? *British Journal of Political Science*, 21 (2), 215–233. doi:10.1017/S0007123400006116

Biedenkopf, K., 2012. *Emissions trading—a transatlantic journey for an idea?* Berlin: Kolleg-Forschergruppe "The Transformative Power of Europe", Working Paper No. 45.

Biedenkopf, K. and Torney, D., 2014. Cooperation on greenhouse gas emissions trading in EU-China climate diplomacy. *In*: E. Reuter and J. Men, eds. *China-EU: green cooperation*. Singapore: World Scientific Publishing, 21–38.

Biedenkopf, K., Bachus, K., and van Eynde, S., 2019. Environmental, climate and social leadership of small enterprises: Fairphone's step-by-step approach. *Environmental Politics*, 28 (1).

Blondel, J., 2014. What have we learned? *In*: R.A.W. Rhodes and P. 'T Hart, eds. *The Oxford handbook of political leadership*. Oxford: Oxford University Press, 705–718.

Carbon Disclosure Project, 2016. *Out of the starting blocks: tracking progress on corporate climate action*. London: Carbon Disclosure Project.

Carmin, J. and VanDeveer, S.D., eds., 2004. *EU enlargement and the environment: institutional change and environmental policy in Central and Eastern Europe*. Abingdon: Routledge.

Carter, N. and Childs, M., 2018. Friends of the earth as a policy entrepreneur: 'the big ask' campaign for a UK climate change act. *Environmental Politics*, 27 (6). doi:10.1080/09644016.2017.1368151

Checkel, J.T., 2005. International institutions and socialization in Europe: introduction and framework. *International Organization*, 59 (4), 801–826. doi:10.1017/S0020818305050289

Cortell, A.P., and Davis Jr., J.W., 2000. Understanding the domestic impact of international norms: a research agenda. *International Studies Review*, 2 (1), 65–87. doi:10.1111/1521-9488.00184

Council of the European Union, 2017. *Climate finance: EU and member states' contributions up to €20.2 billion in 2016* [online]. Available from: http://www.consilium.europa.eu/en/press/press-releases/2017/10/17/climate-finance-eu/ [Accessed 9 February 2018].

Damro, C., 2006. EU-UN environmental relations: shared competence and effective multilateralism. In K.V. Laatikainen and K.E. Smith, eds. *The European Union at the United Nations: intersecting multilateralisms*. Basingstoke: Palgrave Macmillan, 175–192.

Damro, C., 2012. Market power Europe. *Journal of European Public Policy*, 19 (5), 682–699. doi:10.1080/13501763.2011.646779

Dolowitz, D. and Marsh, D., 1996. Who learns what from whom: a review of the policy transfer literature. *Political Studies*, 44 (2), 343–357. doi:10.1111/j.1467-9248.1996.tb00334.x

Dorsch, M.J. and Flachsland, C., 2017. A polycentric approach to global climate governance. *Global Environmental Politics*, 17 (2), 45–64. doi:10.1162/GLEP_a_00400

Dupont, C., Oberthür, S., and Biedenkopf, K., 2018. Climate change: adapting to evolving internal and external dynamics. *In*: C. Adelle, K. Biedenkopf, and D. Torney, eds. *European Union external environmental policy: rules, regulation and governance beyond borders*. Basingstoke: Palgrave, 105–124.

European Council, 2007. *Brussels European Council, 8-9 March 2007: presidency conclusions*. Brussels: Council of the European Union.

European Council, 2014. *European Council (23 and 24 October 2014)–conclusions*. Brussels: European Council.

Gupta, J. and Grubb, M., eds, 2000. *Climate change and European leadership: a sustainable role for Europe?* Dordrecht: Kluwer Academic.

Jänicke, M., 2010. German climate change policy: political and economic leadership. *In*: R.K.W. Wurzel and J. Connelly, eds. *The European Union as a leader in international climate change politics*. London: Routledge, 129–146.

Jänicke, M. and Wurzel, R., 2019. Leadership and lesson-drawing in the European Union's multilevel climate governance system. *Environmental Politics*, 28 (1).

Jordan, A., et al., 2015. Emergence of polycentric climate governance and its future prospects. *Nature Climate Change*, 5, 977–982. doi:10.1038/nclimate2725

Karlsson, C., et al., 2012. The legitimacy of leadership in international climate change negotiations. *Ambio*, 41, 46–55. doi:10.1007/s13280-011-0240-7

Kern, K., 2019. Cities as leaders in EU multi-level climate governance? Embedded upscaling of local experiments in Europe. *Environmental Politics*, 28 (1).

Kilian, B. and Elgström, O., 2010. Still a green leader? The European Union's role in international climate negotiations. *Cooperation and Conflict*, 45 (3), 255–273. doi:10.1177/0010836710377392

Liefferink, D., et al., 2018. Leaders and pioneers in polycentric climate governance. *In*: A. Jordan, ed. *Governing climate change: polycentricity in action?* Cambridge: Cambridge University Press, 135–151.

Liefferink, D. and Wurzel, R.K.W., 2017. Environmental leaders and pioneers: agents of change? *Journal of European Public Policy*, 24 (7), 651–668. doi:10.1080/13501763.2016.1161657

Lindenthal, A., 2014. Aviation and climate protection: EU leadership within the international civil aviation organization. *Environmental Politics*, 23 (6), 1064–1081. doi:10.1080/09644016.2014.913873

March, J.G. and Olsen, J.P., 1998. The institutional dynamics of international political orders. *International Organization*, 52 (4), 943–969. doi:10.1162/002081898550699

Meyer, J.-H., 2011. *Appropriating the environment: how the European institutions received the novel idea of the environment and made it their own*. Berlin: Kolleg-Forschergruppe "The Transformative Power of Europe", Working Paper No. 31.

Oberthür, S. and Dupont, C., 2010. The council, the European Council and international climate policy: from symbolic leadership to leadership by example. *In*: R. K.W. Wurzel and J. Connelly, eds. *The European Union as a leader in international climate change politics*. London: Routledge, 74–91.

Oberthür, S. and Roche Kelly, C., 2008. EU leadership in international climate policy: achievements and challenges. *The International Spectator*, 43 (3), 35–50. doi:10.1080/03932720802280594

Ostrom, E., 2010. Polycentric systems for coping with collective action and global environmental change. *Global Environmental Change*, 20 (4), 550–557. doi:10.1016/j.gloenvcha.2010.07.004

Ostrom, E., 2012. Nested externalities and polycentric institutions: must we wait for global solutions to climate change before taking actions at other scales? *Economic Theory*, 49 (2), 353–369. doi:10.1007/s00199-010-0558-6

Parker, C.F., et al., 2012. Fragmented climate change leadership: making sense of the ambiguous outcome of COP-15. *Environmental Politics*, 21 (2), 268–286. doi:10.1080/09644016.2012.651903

Parker, C.F. and Karlsson, C., 2017. The European Union as a global climate leader: confronting aspiration with evidence. *International Environmental Agreements*, 17 (4), 445–461. doi:10.1007/s10784-016-9327-8

Parker, C.F. and Karlsson, C., 2018. The UN climate change negotiations and the role of the United States: assessing American leadership from Copenhagen to Paris. *Environmental Politics*, 27 (3), 519–540. doi:10.1080/09644016.2018.1442388

Parker, C.F., Karlsson, C., and Hjerpe, M., 2017. Assessing the European Union's global climate change leadership: from Copenhagen to the Paris Agreement. *Journal of European Integration*, 39 (2), 239–252. doi:10.1080/07036337.2016.1275608

Postnikov, E., 2018. Environmental instruments in trade agreements: pushing the limits of the dialogue approach. *In*: C. Adelle, K. Biedenkopf, and D. Torney, eds. *European Union external environmental policy: rules, regulation and governance beyond borders*. Basingstoke: Palgrave, 59–80.

Rayner, T. and Jordan, A., 2010. The United Kingdom: a paradoxical leader? *In*: R. K.W. Wurzel and J. Connelly, eds. *The European Union as a leader in international climate change politics*. London: Routledge, 95–111.

Rayner, T. and Jordan, A., 2013. The European Union: the polycentric climate policy leader? *WIREs Climate Change*, 4 (2), 75–90. doi:10.1002/wcc.205

Risse-Kappen, T., 1994. Ideas do not flow freely: transnational coalitions, domestic structures, and the end of the cold war. *International Organization*, 48 (2), 185–214. doi:10.1017/S0020818300028162

Rose, R., 1993. What is lesson-drawing. *Journal of Public Policy*, 11 (1), 3–30. doi:10.1017/S0143814X00004918

Sabel, C.F. and Victor, D.G., 2016. *Making the Paris process more effective: a new approach to policy coordination on global climate change*. Muscatine, IA: The Stanley Foundation.

Schimmelfennig, F. and Sedelmeier, U., 2004. Governance by conditionality: EU rule transfer to the candidate countries of Central and Eastern Europe. *Journal of European Public Policy*, 11 (4), 661–679. doi:10.1080/1350176042000248089

Schreurs, M.A. and Tiberghien, Y., 2007. Multi-level reinforcement: explaining European Union leadership in climate change mitigation. *Global Environmental Politics*, 7 (4), 19–45. doi:10.1162/glep.2007.7.4.19

Skovgaard, J., 2013. The limits of entrapment: the negotiations on EU reduction targets, 2007-11. *Journal of Common Market Studies*, 51 (6), 1141–1157. doi:10.1111/jcms.12069

Tews, K., Busch, P.-O., and Jörgens, H., 2003. The diffusion of new environmental policy instruments. *European Journal of Political Research*, 42 (2), 569–600. doi:10.1111/1475-6765.00096

Torney, D., 2015a. Bilateral climate cooperation: the EU's relations with China and India. *Global Environmental Politics*, 15 (1), 105–122. doi:10.1162/GLEP_a_00274

Torney, D., 2015b. *European climate leadership in question: policies toward China and India*. Cambridge, MA: MIT Press.

Torney, D., 2017. If at first you don't succeed: the development of climate change legislation in Ireland. *Irish Political Studies*, 32 (2), 247–267. doi:10.1080/07907184.2017.1299134

Torney, D. and Gippner, O., 2018. China: deepening cooperation on climate and environmental governance. *In*: C. Adelle, K. Biedenkopf, and D. Torney, eds. *European Union external environmental policy: rules, regulation and governance beyond borders*. Basingstoke: Palgrave, 275–296.

Torney, D. and Cross, M.K.D., 2018. Climate and environmental diplomacy: building coalitions through persuasion. *In*: C. Adelle, K. Biedenkopf, and D. Torney, eds. *European Union external environmental policy: rules, regulation and governance beyond borders*. Basingstoke: Palgrave, 39–58.

van Schaik, L. and Schunz, S., 2012. Explaining EU activism and impact in global climate politics: is the union a norm- or interest-driven actor? *Journal of Common Market Studies*, 50 (1), 169–186. doi:10.1111/j.1468-5965.2011.02214.x

Vogel, D., 1995. *Trading up: consumer and environmental regulation in a global economy*. Cambridge, MA: Harvard University Press.

Vogler, J., 1999. The European Union as an actor in international environmental politics. *Environmental Politics*, 8 (3), 24–48. doi:10.1080/09644019908414478

Wurzel, R., Liefferink, D., and Torney, D., 2019. Pioneers, leaders and followers in multilevel and polycentric climate governance. *Environmental Politics*, 28 (1).

Youngs, R., 2001. *The European Union and the promotion of democracy: Europe's Mediterranean and Asian policies*. Oxford: Oxford University Press.

Index

Note: **Bold** page numbers refer to tables; *italic* page numbers refer to figures and page numbers followed by "n" denote endnotes.

Aarhus 32
Action Aid Netherlands 52
Al-Falih, Khalid A. 90
altruists 69
American Petroleum Institute (API) 101n4
Andresen, S. 45
associated British Ports (ABP) 158

Baden-Württemberg 29
Bavaria 30
Berkhout, F. 37
Biedenkopf, K. 12, 13
bio-energy village 30
Blair, Tony 27, 180
bottom-up governance 105
Bremerhaven 149, 153, 155, 157, 161
burden sharing agreement 25
Burns, J.M. 9, 11
business 5, 7, 9

C40 Clean Bus Declaration 31
California electronic waste recycling Act 47
carbon capture and storage (CCS) 111–112, 114
cause–effect relations 10
cities' formal institutional powers 151
citizen cooperatives 32
clean energy transformation 27
clean power supply 30
Climate Change Act 27
climate change politics 6
climate coalition 79
climate follower 170–172
climate followership: pathways, materialise 172–174; *see also* followership
climate governance: actors 5; innovations **156**; 'leaderless leader' paradox 32–37; literature 2; multilevel climate governance, EU support 35–37
climate laggard 87–101; *see also* laggards
climate leaders 65, 87–101; cities 31; regions 31; villages 30, 31
climate leadership 24, 66; Coop and Migros climate behaviour 77–78; dynamics 34; motives and capacities 78–79; regulatory framework 80–82; social interactions 79–80; theory building, pathway 78–82
climate leadership dimensions 25–28; Denmark 27–28; Germany 25–26; United Kingdom 26–27
climate lighthouses 153
climate policy integration (CPI) 34
climate pushers 64–83; Coop and Migros 77
CO_2 reduction 31
cognitive leadership 10, 11, 14, 23, 48, 51, 52, 54–56, 96, 151, 168, 173
conditional pushers 8
Conference of the Parties (COP) 91, 95
conflict minerals 46
conflict resolution 13
constructive pushers 8
contemporary climate governance scholarship 2
convergence 168
Copenhagen climate conference (COP15) 179
Copenhagen climate negotiations 117
Cornish village of Delabole 31
corporate actors 6, 70, 82; in polycentric climate governance 82–83
corporate climate leadership: cases studies 73–74; case studies 71–72; Coop and Migros 73–74; data sources and analysis

73; exploratory framework on *70*; food retailers 64–83; framework 69–71; global food sector and climate 71; mechanism of *81*; Swiss food retailers 72
corporate climate strategies 66, 105; discursive face 76; internal face 75–76; legally binding commitments, overcompliance 74–75; political face 76; and public climate policy 74–75; typology for analysing 66–69; voluntary agreement 74
corporate leadership 105
Corporate Social Responsibility programmes 47
corporatism 7
Council of European Municipalities and Regions (CEMR) 134
Covenant of Mayors (CoM) 127, 137, 138, 140, 141
Covenant of Mayors for Climate & Energy 137

Damro, C. 177
Darnall, N. 65
Denmark 25, 27–28
diffusion, local experiments upscaling 128, 129
directional leadership 11
discrete activists 69
discursive face 67–68
distinct leadership 46
Dodd-Franck Act 47
Dorsch, M.J. 3
Dupuis, J. 12

early followers 23–28
economic power 14, 23
eco-premiums 67
eco-profit programme 30
Eikeland, P.O. 88
electric power industry 118
embedded upscaling 136–138
emissions trading scheme (ETS) 26–27, 177
e-mobility 30
end-of-life treatment actors 51, 55, 56
Energy City brand 155
energy consumption 46
Energy Savings Law, Germany 135
Energy Savings Ordinance, Germany 135
energy transition 11
Energy Union 36
entrepreneurial leadership 9–11, 23, 49–51, 53, 54, 95, 151, 168, 173
environmental ambitions 7, 8

environmental capacity literature 7
environmental governance literature 1
environmental innovation 65
environmental lawsuits, Japan 24
environmental leaders 6, 22
environmental NGOs (ENGOs) 5
environmental policy integration (EPI) 34
Environmental Protection Agency (EPA) 24
epistemic communities 6
EU emissions trading system (ETS) 106, 107, 109, 118–120; 2008 reform, responses 112–115; initial corporate responses 110–112; observed strategic responses 115; proactive responses, factors 116–118; 'reactive' and 'proactive' responses models 115; reforms 110
EU multilevel climate governance: analysis, research design and methods, character 127–128; embedded upscaling 136–138; hierarchical upscaling 134–136; horizontal upscaling 129–133; local experiments, upscaling 128–129; subnational leadership 28–32; upscaling and networking 138–140; vertical upscaling 133–134; *see also specific entries*
Eurelectric's proactive political response 115
European Energy Award (EEA) 7, 153
European Fund for Strategic Investment (EFSI) 36
European Green Capital Awards 7, 35
European Investment Bank (EIB) 36
Europeanisation 37
European Structural and Investment Funds (ESIF) 35
European Union (EU) climate policy 22
Europia 115, 116
exemplary leadership 11, 23, 48–52, 54, 56, 96, 152, 168
exemplary pioneership 11
expansion, local experiments upscaling 128

Fairphone 2 52, 53
Fairphone's step-by-step approach 43–57; leadership types and **55**
Fischer, Joschka 29
Flachsland, C. 3
followership 12–15, 174–176; coercion 177; dimensions of **176**; EU leadership and (non-)followership 176–177; incentives 178; learning 179–180; persuasion 180–181
free riders 68

INDEX

Freiburg 29
Fuel Duty Escalator 26
Fuels Europe 101n4, 114
functional networking 141
functional networks 139

German renewable energy law
  (*Erneuerbare Energien Gesetz* (EEG)) 28
Germany 25–26, 126
Gerring, J. 72
Ghanaian Environmental Protection
  Agency 55
Global Climate Coalition (GCC) 87
global decarbonisation 2
'Green-Black' (Greens-CDU) 29
green economy 157–160
greenhouse gas emission (GHGE) 25–27,
  36, 88, 89, 147, 153
Greenhouse Gas (GHG) Protocol
  methodology 73
'green' lead markets 7
Green Party 25, 29
'Green-Red' (Greens-SPD) 29
Green Troika 5
greenwasher 67
Grubb, M. 11
Gupta, J. 11

Hanseatic League 127
Hayward, J. 23
Hesse Energy Future Law 29
hidden lobbyists 69
hierarchical governance 135
hierarchical upscaling 134–136
horizontal Europeanisation 37
horizontal upscaling 129–133
Hull 149, 154, 155, 160, 161
Hull City Council 159
Humber Local Enterprise Partnership
  (LEP) 159

ICLEI 141n1
Intended Nationally Determined
  Contributions (INDCs) 174, 178
intentional exemplary leaders 152
intentional exemplary leadership 11
interactive learning, dynamic system 34
interconnectedness 12
Intergovernmental Panel on Climate
  Change (IPCC) 87
international demonstration effect 24
IPIECA 101n4

joint decision traps 34
Jonas, A.E.G. 148

Jordan, A. 2, 26, 169
Jörgens, H. 37
Jühnde 30

Kaelble, H. 149
Kern, K. 171
Ki-moon, Ban 125
Klemmer, P. 38n3
Kurs Klimastadt 153
Kyoto Protocol 2, 25, 75, 111, 169, 178, 182

laggards 68, 104–121
Land Bremen 154, 157
leader–follower relationships 12, 13, 16,
  167, 169–171
leader-laggard typology 89
'leaderless leader' paradox 23, 32–37
leadership 3, 5–8, 14; follower groups **50**;
  studies 1; types 8, **50**; *see also* climate
  leadership
learning-by-doing 3
Le Galès, P. 149, 163
lesson-drawing 22, 23, 33, 34
Liefferink, D. 8, 25, 38n2, 44, 48, 49, 56,
  57, 66, 67, 89, 91–92, 94–96, 98, 99, 148,
  150, 162, 163, 168, 171–173
Lima-Paris Action Agenda 169
local experiments, upscaling 128–129
Locke, R.M. 44, 46
London 31
low-carbon leaders 104–121

Maastricht Treaty 33
Malnes, R. 16
March, J.G. 170
'Market Power Europe' 177
market responses 106, 111
Market Stability Reserve 120n8
Master Plan Active Climate Policy 153
Masterplankommunen (MPK) 137
meta-networking 141
Montreal Protocol 68
Morrison, T.H. 9, 12
multi-impulse-hypothesis 38n3
multi-impulse mechanism 33
multilateral UN process 2
multinational companies 43
Munich 30

national climate leaders 24–25
National Environmental Policy Plan
  (NEPP) 25
national leaders 23–28, 37; climate
  leadership dimensions 25–28; national
  climate leaders 24–25

national oil companies (NOCs) 100n2
Niemann, H. 30
Non-Fossil Fuel Obligation 26
non-state actors 8, 10, 11, 13, 15, 45, 57, 88

Oberthür, S. 2, 9
Odyssee Klima 155
Offshore Terminal Bremerhaven (OTB) 157, 158
Oil and Gas Climate Initiatives (OGCI) 88–93, 95–100; position, determining 93–94; potential climate leader 97–98
oil and gas sector 87–101; conditions 94–95; internal and external, positions 92–93; leadership styles 97; leadership types 96–97; methods 89–90; OGCI 90–91; positions, conditions, styles and strategies 91; styles and strategies 95–96
oil and power industries: analytical point of departure 106–109; corporate 'proactive' responses, factors 108–109; EU emissions trading 104–121; 'proactive' corporate responses 107–108; reactive corporate responses 106–107
Olsen, J.P. 170
Ostrom, Elinor 9, 136, 169

Paris agreement 2, 33, 98, 114, 126
peer-to-peer learning 38
pioneers 6, 69
pioneership 3, 5–8
policy convergence 23
policy diffusion 168
policy transfer and learning 168
political 'face' 68
polycentric concepts 10
polycentric governance approaches 2, 3, 13, 15, 16
polycentricity 2–4, 11; proponents of 3
polycentric system 12
Porter, M. 107, 109
Porter hypothesis 65, 66, 70, 80
Postnikov, E. 178
potential environmental leaders 4
potential laggards 106–107
power 9
public actors 13
public climate policy 74–75
public governance actors 170

Radaelli, C. 23
Räthzel, N. 5
Rayner, T. 26

reactive corporate responses 106–107
Reckien, D. 131
regional actors 11
regulatory framework 69
renewable energy 32, 113
'revolutionary' leadership 11
Rhodes, R.A.W. 12
Rio Earth Summit 125
Röhl, K.-H. 149
Rose, R. 22
RWE 111, 113, 119

Sabel, C.F. 169
Samel, H.M. 46
Schlossberg, T. 31
Schreurs, M.A. 4, 32
Schrijf-Schrijf, consultancy 52
Schröder, C. 149
Schweizer, R. 12
scientists 6
Seawright, J. 72
self-organisation 10
self-regulation 5
Shell 111, 112
Siemens-ABP Green Port Hull development 159
single country studies 4
single European market (SEM) 23
Singleton, B.E. 9
Skjærseth, J.B. 88
Skodvin, T. 45
small and medium enterprises (SMEs) 45
small enterprises 43–57
smartphones: battery management 47; environmental, climate and social impact 46–48; environmental, climate and social leadership 48–51; life cycle 48–51
Smart Specialisation Platform 35
Social Democratic-Liberal coalition 24
social enterprises 13, 44, 49, 53
social entrepreneurs 44
social interactions 70
social leadership 51–56
societal actors 3
societal participation strategies 155–156
Solorio, I. 37
state actors 10, 11, 23
state-centred approaches 4
structural leadership 8, 9, 14, 23, 27, 48, 95, 168
structurally disadvantaged cities (SDCs) 147, 148, 162, 163; climate governance

in 152–160; climate pioneership and leadership 150–152; green economy 157–160; multi-level, polycentric or place-specific, governance 160–161; small-to-medium-sized 148–150, 161; societal participation strategies 155–156; urban climate governance in 160–161; urban climate strategy 152–155
subnational actors 11, 15
subnational leadership 28–32; Denmark 31–32; Germany 29–30; United Kingdom 31
substantive leaders 172
substate actors 23
supply chain leadership 45
supranational actors 9, 10
Sweden 25
Swedish Confederation of Professional Employees (TCO) sustainability certification 52
symbolic leaders 8, 64–83, 172
systemic opportunity structures 37

territorial networking 139, 141
'T Hart, P. 12
Thatcher, Margaret 26
Tiberghien, Y. 4, 32
top-down climate governance approach 2
top-down governance 105
trade union movements 5
trade unions 5
transactional leadership 11–12, 23, 96, 152
transformation, local experiments upscaling 128, 129
transformational leadership 96, 152, 175

transformative leadership 23
transnational networks 23
true ownership 54
The Truly Disadvantaged 148

Underdal, A. 6, 16
UN Framework Convention on Climate Change (UNFCCC) 64, 93, 94
unintentional example-setting 11
United Kingdom 25, 26–27
UN Rio conference 24
urban climate strategy 152–155
Uzzell, D. 5

Van der Linde, C. 107, 109
Vattenfall 113, 119
vertical upscaling 133–134
Victor, D.G. 169
voluntary agreements 5

Waag Society 52
Wildpoldsried 30
Wilson, W.J. 148
Windenergie Agentur Bremerhaven (WAB) 157
World Bank 128
World Business Council on Sustainable Development (WBCSD) 73
World Resources Institutes (WRI) 73
World Trade Organization (WTO) 178
Wurzel, R.K.W. 8, 13, 25, 38n2, 44, 48, 49, 56, 57, 66, 67, 89, 91–92, 94–96, 98, 99, 148, 150, 162, 163, 168, 171–173

Young, O.R. 8, 16

Milton Keynes UK
Ingram Content Group UK Ltd.
UKHW022145300824
447584UK00002B/12

9 780367 467593